Convergent Journalism

PETER LANG
New York • Washington, D.C./Baltimore • Bern
Frankfurt am Main • Berlin • Brussels • Vienna • Oxford

Stephen Quinn

Convergent Journalism

The Fundamentals of Multimedia Reporting

PETER LANG
New York • Washington, D.C./Baltimore • Bern
Frankfurt am Main • Berlin • Brussels • Vienna • Oxford

Library of Congress Cataloging-in-Publication Data
Quinn, Stephen.
Convergent journalism: the fundamentals of multimedia reporting / Stephen Quinn.
p. cm.
Includes bibliographical references and index.
1. Journalism. 2. Online journalism. 3. Broadcast journalism.
4. Convergence (Telecommunication). I. Title.
PN4731.Q56   070.4—dc22   2005010677
ISBN 978-0-8204-7452-6

Bibliographic information published by **Die Deutsche Bibliothek**.
Die Deutsche Bibliothek lists this publication in the "Deutsche Nationalbibliografie"; detailed bibliographic data is available on the Internet at http://dnb.ddb.de/.

Cover design by Joni Holst

© 2005, 2009 Peter Lang Publishing, Inc., New York
29 Broadway, 18th floor, New York, NY 10006
www.peterlang.com

All rights reserved.
Reprint or reproduction, even partially, in all forms such as microfilm, xerography, microfiche, microcard, and offset strictly prohibited.

In loving memory of my father,
Ronald Alfred Quinn,
26 November 1922 to 9 July 2004

Contents

	List of Illustrations	ix
	Acknowledgments	xi
1	The Emergence of Convergence	1
2	Why and How Convergence Is Emerging	27
3	Business and Revenue Models for Convergence	61
4	Convergent Journalism and Multi-Media Storytelling	85
5	Case Studies of Convergence	105
6	Technology and Convergence	13
7	The Smart Newsroom: Knowledge Management and Convergence	153
8	Convergence and the Future of Journalism	177
9	Implementing Convergence in the Newsroom	201
	Glossary	223
	References	231
	Index	247

Illustrations

1	The CLTV news desk in the newsroom of *The Chicago Tribune*	9
2	The cluttered and busy newsroom of BBC Interactive in London	37
3	The newsroom of the *Orlando Sentinel* with the TV newsroom in the distance	49
4	Julie Nichols, Newsplex project director, confers with news resourcer Geoff LoCicero	95
5	Reporters at the *Lawrence Journal-World* enjoy the airy feel of their converged newsroom	109
6	The continuous news desk at *The New York Times* is the nerve center for breaking news	122
7	Newsplex founder and Ifra director of publications Kerry Northrup provides feedback to trainees using the news wall	146
8	Training remains an ongoing need for journalists	149
9	The training newsroom of *The Columbia Missourian* at the University of Missouri	161
10	Much tacit knowledge transfer takes place when journalists work in teams, such as at *The Chicago Tribune*	167
11	An aerial view of the multi-media desk at the Tampa News Center, taken early in the day	186
12	The multi-media desk at the *Lawrence Journal-World* with its polished wooden floors	217

Acknowledgments

Many people helped with the research for this book. It would occupy too many pages to list all of them. But I would like in particular to express my thanks to the people who were especially helpful—who listened patiently and answered my questions, however naïve those questions were. I have listed those people in alphabetical order of their employers. Thank you all. And thanks to Kathleen DuVall, who did a fine job copyediting the manuscript. Any errors are mine alone.

Special acknowledgment must go to Kerry Northrup at the Newsplex for his generosity of spirit and advice, despite a hugely busy schedule. Kerry probably knows more about convergence than any one person in the world.

Here I would like to thank all the people who contributed their time:

Journalists and Managers

American Press Institute: Andrew Nachison
Archronica Architects in New York: Saf Fahim
Associated Newspapers in London: Allan Marshall
BBC-TV: Paul Brannan and Paul Myles
Danish Center for Journalism Education: Jan Larsen
Fairfax Holdings in Australia: Fred Hilmer
Ifra Asia in Singapore: Elaine Wong
Ifra's *newspaper techniques* in Germany: Dean Roper
    and Brian Veseling
Ming Pao Group in Hong Kong: Paul Cheung
NewsLink Indiana: Terry Heifetz and Wright Bryan
Newsplex at the University of South Carolina: Kerry Northrup,
    Julie Nichols, and Geoff LoCicero
NordJyske in Denmark: Ulrik Haagerup
Orlando Sentinel: Keith Wheeler

Ottawa Citizen: Chris Cobb
San Francisco Chronicle: Kenn Altine and Joe Brown
Singapore Press Holdings: Thomas Wee
St. Petersburg Times: Deb Wolfe
Tampa Tribune and WFLA: Forrest Carr, Ken Knight, Victoria Lim, and Gil Thelen
The Age in Australia: Jacqui Cheng, Colin McKinnon, and Hugh Martin
The Chicago Tribune: Mark Hinojosa and Howard Tyner
The New York Times: Stephen Miller
The Times in London: George Brock
The World Company in Lawrence, Kansas: Rob Curley
Turun Sanomat in Finland: Ari Valjakka

Academics

Ball State University: Larry Dailey, Dr. Vince Filak, Bob Papper and my graduate assistant, Kintija Eigmina
Indiana University at Purdue: Dr. Edgar Huang
Northwestern University: Rich Gordon
The University of Iowa: Dr. Jane Singer
The University of Kansas: James Gentry and Rick Musser
The University of Minnesota: Nora Paul
The University of Southern California: Larry Pryor
The University of South Carolina: Charles Bierbauer
Zayed University in the United Arab Emirates: Dr. Tim Walters

1 The Emergence of Convergence

Convergence is happening around the world. This chapter looks at the evolution of this phenomenon—also known as multiple-platform publishing or integrated journalism—around the world, and attempts to define the concept. It also considers the key question of whether editorial managers are adopting it to save money or to do better journalism, and concludes that the two issues go hand in hand. The forces driving convergence are described briefly, and the chapter concludes with an overview of how to foster convergence. The key elements of the chapter include:

* Defining convergence
* Business models and convergence
* Convergence and digital tools
* Forces driving convergence
* Fostering convergence
* The key role of managers

Convergence is a likely destination for the news media in many parts of the world, though the duration of the journey will vary from country to country. Some media organizations are eagerly embracing the concept, seeing it as a way to deal with an uncertain future. Others are hanging back, waiting to see what evolves. The chairman of the New York Times Company and publisher of *The New York Times,* Arthur O. Sulzberger Jr., is a leader in the first group. In February 2004 he told a conference at Northwestern University that convergence was "the future" for the media. He described how his company had been acquiring other media outlets such as the Discovery Channel to allow *Times* journalists to tell stories in print, online, and on television. "Broadband is bringing us all together," Sulzberger said. "We have to do it in papers, digitally and on TV. You can combine all three elements. News is a 24–7 operation, and if you don't have the journalistic muscles in all three [platforms], you can't succeed in broadband."

Sulzberger described the process as "a hell of a challenge" (quoted in Damewood 2004). The next month the chief operating officer of one of the biggest media groups in Canada echoed Sulzberger's comments. CanWest Global Communications Corporation's Rick Camilleri told a conference in Toronto that convergence was the only viable business model for media companies in the digital age. Time Warner's "vertical" convergence model, which tried to combine content and distribution through a merger, was bound to fail, Camilleri said. CanWest was forging ahead with its plans to launch "horizontal, branded content exploited over different media platforms" (Wells 2003: 1). As of late 2004, CanWest owned 11 major Canadian daily newspapers, including *The Gazette* and the *National Post*. The company's Global Television Network also owned 11 television stations that reached 94 percent of English-speaking Canada, plus two television networks in New Zealand and others in Australia and Ireland.

Elsewhere in the world, news organizations have been embracing convergence at different levels, often faster than in the US. In 2001 Juan Antonio Giner, founder of the Innovation International media consulting group, wrote that seven out of 10 newspaper executives said their reporters had formal duties in at least one other medium apart from the newspaper (2001a: 28). Newspapers were becoming "24 hour information engines," just as broadcast organizations like CNN had become 24-hour news providers. "Media diversification is the past. Digital convergence is the present. Multi-media integration is the future," Giner wrote in the online edition of *Ideas*, the journal of the International Newspaper Marketing Association (INMA). Earl Wilkinson, INMA's executive director, noted after attending a newspaper conference in Singapore that: "The major newspaper companies worldwide have accepted the multimedia, brand-oriented future for newspapers" (quoted in Giner 2001b). A year later, Martha Stone, at the time a senior consultant for Innovation International, wrote that in nearly every country on each continent mono-media companies were "transforming into multi-media companies, integrating editorial side operations from print, web and broadcast divisions." The benefits of convergence were "overwhelming," she said. Stone noted that 73 percent of the members of the World Association of Newspapers (WAN) had reported some form of convergence emerging at their companies (2002b: 1).

Media companies in Southeast Asia and Scandinavia have embraced convergence most widely as of mid-2004. InSoutheast Asia the leaders included Star Publications in Kuala Lumpur, Malaysia; the Nation group in Thailand; *JoongAng Ilbo* and the Maeil Business Group in South Korea; the Singapore Press Holdings group, which publishes the prestigious *Straits Times* newspaper; and the Ming Pao Group in Hong Kong. Scandinavian media groups were especially advanced. *Aftonbladet* and the Bonnier group were pioneers in Sweden. Norway's leader was the *Aftenposten* and *NordJyske* was their Danish counterpart. The Turun Sanomat Group in southwest Finland was one of the world's leaders in multiple-platform publishing. Editor-in-chief Ari Valjakka said the key issue in Finland was people's time: individuals spent an average of 7.5 hours a day in media-related activities. "The division of time between all possible information channels is fierce and that's why you need to be involved in more than one [medium]" (Valjakka 2002b).

In Europe, the Bertelsmann group—Europe's biggest media company—was a pioneer in media convergence in Germany (though it cut back its involvement after running into financial problems in 2002). In the UK the leaders were the *Financial Times, The Guardian,* and the BBC. In Spain, the Marca Group captured 62 percent of the daily sports market through a combination of the daily newspaper (circulation 564,000) and a huge website that offered plenty of multi-media content. Other Spanish leaders were *Recoletos, El Mundo,* and Grupo Correo.

In the Middle East, strategic alliances and mergers within the Arab media were expected to strengthen some companies as the media expanded significantly there. One of these expansions provided an early example of convergence. In Beirut, the Lebanese Broadcasting Corporation (LBC) and the London-based Arabic newspaper *Al-Hayat* were spending $12 million a year in a joint venture in which the newspaper's 69 correspondents supplied news for LBC International's bulletins, starting in 2002. LBC's managing editor, Salameh Nemett, said if the venture succeeded another 24-hour news channel could emerge. Saudi money was behind this partnership (Khalaf 2003: 12).

In South America, Juan Antonio Giner listed the leaders as *Clarin* in Argentina; the Reforma Group and Televisa in Mexico; *O Globo*, the O Estado de S. Paulo Group and RBS in Brazil; *El Universal* in Venezuela; *El Caribe* in the Dominican Republic; *El Nuevo Dia* in Puerto Rico; *Telefuturo* in Paraguay; *El Tiempo* in Colombia; and *Medcom* in Panama. Giner calculated in 2001 that more than 100 newspaper companies around the world were on their way to full multi-media integration (Giner 2001b). A little over three years later the number had jumped to about 475 (Northrup 2004a).

The Bell Globe Media group led the convergence charge in Canada. It owned the national daily *The Globe and Mail,* and a television news service, The Business Report. In the US, the pioneers tended to be grouped in Florida: the *Tampa Tribune,* the *Orlando Sentinel,* the *Sarasota Herald-Tribune* and *Florida Today.* Other leaders were *The Washington Post, The Los Angeles Times,* and *The Chicago Tribune.* The American Press Institute has maintained a convergence tracker on its website since spring of 2002. It was the brainchild of API's media center director Andrew Nachison and Jimmy Gentry at the University of Kansas. Gentry's multi-media management graduate class assembled the information for the initial database, and a former student updates it in conjunction with API staff.

As of late 2004 the tracker showed convergence happening in 33 of the 48 mainland states (Gentry 2004b, API 2004). Howard Finberg of the Poynter Institute calculated that about 100 of the 1,457 daily newspapers in the US had embraced convergence as of early 2004. Chapter 5 details case studies of successful converged media groups around the world. Early in 2004 the Project for Excellence in Journalism, an institute affiliated with Columbia University's Graduate School of Journalism, published a study of the state of the American news media. The Pew Charitable Trusts funded the study, which identified eight media trends. Convergence was one of them: "Convergence seems more inevitable and potentially less threatening to journalists than it may have seemed a few years ago." Online journalism appeared to be leading the move to convergence, the study said. A growing number of news outlets were chasing relatively static or shrinking audiences. "One result of this," the study noted, "is that most sectors of the news media are losing audience. That audience decline, in turn, is putting pressures on revenues and profits, which leads to a cascade of other implications" (Project for Excellence 2004).

Defining Convergence

Convergence has as many definitions as the number of people who try to define or practice it. Keynote speakers at a November 2002 conference devoted to defining convergence in the South Carolina capital, Columbia, found it difficult to agree. Charles Bierbauer, dean of the College of Mass Communications and Information Studies at the university, introduced the conference. Two years afterwards, when asked to reflect on the event, he noted that "we had more definitions than presenters (Bierbauer 2004).

The differences of opinion reflected the fact that convergence varies from country to country, company to company, and culture to culture. Cultural factors are paramount in at least two senses of the word: Introducing and managing convergence involves appreciating the specific cultures unique to any organization, and the form of convergence that evolves will be a product of those unique cultures. John Haile was the pioneering editor who introduced convergence at the *Orlando Sentinel:* "The big thing I try to emphasize with anyone looking at how to practice journalism across multiple media is the critical need to address the culture of the organization. Unfortunately, very few news organizations ever stop to deal with that, choosing instead to focus on the relationships, newsroom layouts, titles, etc." Haile said success depended on having journalists who could "think multiple media" and who were comfortable working in or with various media (Haile 2004).

A wide set of variables such as legislation, technology, audiences, and the society into which convergence is introduced influence the extent to which convergence is possible, and can also dictate the form that convergence takes. The next chapter discusses these and other factors. Some presenters at the Columbia conference such as Paul Horrocks, editor-in-chief of the *Manchester Evening News,* part of the Guardian Media Group in the UK, bluntly asserted that convergence was about survival. Horrocks said multiple-platform publishing offered many advantages in the crowded advertising and media markets common in most parts of the UK. "It is [about] delivering a product that we know the customers will want." The Manchester group started on the multi-platform road in mid-2001. Horrocks said media organizations intending to introduce convergence needed a top-down commitment. "Journalists, by their nature, don't like change. We have to convince them that we have to serve the customer to retain our jobs" (2002).

A year later Horrocks told a publishing conference that newspapers had to reinvent themselves to be more competitive and to satisfy consumers eager to receive information by different channels (Pascual 2003a: 35). Michael Aeria, at the time deputy publisher of the Star Publications Group in Malaysia, similarly said convergence was an opportunity for his company to reach multiple audiences (2002). Gil Thelen, one of the pioneers of convergence who became publisher of the *Tampa Tribune* in 2003, said multiple-platform delivery should be designed "to help people live [their lives] more easily" (2002a). His newspaper, he said, reflected the fact that many readers led hectic lives and had less time to spend with the publication. "We want to make sure these on-the-run readers can scan the paper quickly and get a good sense of what's important that day. Yet at the same time, we want to make sure that when these busy readers do catch their breath and find time to read the paper more completely, they'll enjoy the context, depth and perspective on the news that only newspapers can provide" (2002b). This remains the defining paradox of the modern newspaper: how can it present its content in such a way that consumers can scan it quickly, yet also offer content that is deep and informative.

A Danish newspaper editor, Ulrik Haagerup, told the Columbia conference that convergence was not about technology but had "everything to do with mindset." Haagerup said convergence was associated with the way that journalists viewed their role in society and how they demonstrated their expertise (Haagerup 2002a). It would be safe to say that convergence had almost as many forms and models as it had advocates. In some parts of the world convergence involved a reporter who was a specialist in a particular subject being able to use all media to distribute the vast amount of information that reporter had gathered. In this situation it was not an example of rehashing one story, but of spreading the reporter's expertise so that his or her knowledge and information appeared in several forms. For example, one reporter at the Spanish sports tabloid *Marca* specialized in writing not about football, but only about the Real Madrid football team. He wrote and broadcast about the team for all forms of media, and was recognized as an authority on the subject. This represented an example of the combination of convergence and knowledge management in the newsroom. Chapter 7 elaborates on the notion of the smart newsroom, which incorporates knowledge management techniques with convergence.

Martha Stone, training director for the Newsplex, said the defining convergence conference of November 2002 helped her and director Kerry Northrup "take the temperature" of the industry and determine where it was and where it needed to go. "We really need to define what we are doing with this multiple-media approach. I think it is clear that there are still a number of definitions [of convergence] and that is just the nature of the beast for now." Stone agreed that the various approaches to convergence around the world were "vastly different" (quoted in Roper 2003a: 30). Larry Pryor, interim editor of the *Online Journalism Review* and *Japan Media Review* and a professor at the Annenberg School for Communication at the University of Southern California, maintained that a definition was vital. A new medium needed a common vocabulary. "If we all have a different concept of what convergence means, we are making it difficult to progress." Pryor suggested that if someone did not offer a meaningful definition, confusion could prevail because "nature abhors a vacuum."

Pryor said a quote from Michael Heim's book *Virtual Realism*, published by Oxford University Press, was apt. "The right name illuminates and enlightens, as we learn from Genesis," says Heim. "Getting the right name goes beyond utility. It touches ethics and civic life. . . . If our language sinks below the clear understanding of things, then we lose the bonds that bring us to speak the same language." Pryor was concerned that allowing the industry to define convergence would produce a simplistic definition with a limited scope. Convergence was more than "corporate conglomeration" or coordination of news production. "I'd prefer to see it defined by people who study the field and do the experimental work. This sounds self-serving, but that would be those of us at universities, although not necessarily in schools of journalism." Pryor called for a joint effort among many disciplines such as engineering, cinema, TV, journalism, business, law, and communications, supported by organizations such as the Poynter Institute and the Pew and Ford foundations, plus contributions from creative people in industry.

This book will not discuss corporate convergence, where big companies merge because of the mutual benefits of amalgamation. Probably the best known of these was the $165 billion AOL-Time Warner merger announced in January 2000. It was touted as convergence because analysts and executives saw advantages in combining Time Warner's content with AOL's networks. Noted Gordon Pitts: "AOL coveted a guaranteed supply of 'content'—entertainment, news, sports and information—to distribute over its powerful Internet portal, as well as access to broadband 'pipes' to ensure that its portal and web applications would have entry to households and offices." AOL chief executive Steve Case wanted to consolidate his company's high stock price by merging with a company with more tangible assets (Pitts 2002: 3–4). Pryor described this form of convergence as an economic model that was "peripheral to journalism," and this book adopts that same approach. *Convergent Journalism* also will not discuss the technological and perhaps utopian form of convergence where many pieces of digital equipment converge in a single box in the living room or study. (For more on this concept read Stewart Brand's *The Media Lab* [1987].)

Pryor said the multi-platform reporting form of convergence was a new form of journalism. For him, websites and other new media forms such as email newsletters, email alerts, wireless feeds, cell phone content, and blogs represented a new medium with "a unique identity and logic of immediacy and transparency." These had some links with traditional media and could draw from them, but convergence also required the ability to produce original content in multi-media forms. This form of journalism was expensive and experimental, Pryor said. "Convergence is what takes place in the newsroom as the editorial staff works together to produce multiple products for multiple platforms to reach a mass audience with interactive content on a 24/7 basis. Anything less is not journalism convergence, in my view" (Pryor 2004).

Definitions of Convergence

"The definition does matter because we are trying to come up with a common vocabulary in this new medium and if we all have a different concept of what convergence means, we are making it difficult to progress."
> Larry Pryor, director of the online journalism program, University of Southern California, Annenberg School for Communication

"There is no single definition of what constitutes convergence because convergence is defined by the media marketplace, and marketplace conditions are different from one location to another. In the broadest sense, convergence is a process whereby media companies break out of their traditional forms and formats to deliver richer news and information services more in concert with the way that consumers are choosing to access and use such resources. It is a response to changes in the media environment brought on by technology and the information economy."
> Kerry Northrup, Ifra director of publications and former Newsplex director

"[Convergence is] serving the customers with multiple channels making use of the basic raw material which in our case is (mostly) news and current information."
> Ari Valjakka, editor-in-chief of Turun Sanomat, Finland

"[Convergence is] a way to serve the community better and to do better journalism by using the most suitable media to tell the story."
> Ulrik Haagerup, editor of Nordjyske, Denmark

"Convergence is not a word I like a lot. It's clunky, it's techno and it's descriptive only in the sense that everything is going digital and in that sense the information streams are becoming common. What I want to focus on is building the multi-media news and information organization and so my definition is a company that operates on multiple platforms to both serve the public interest in terms of news and gains market share in the process."
> Gil Thelen, publisher of the Tampa Tribune, Florida, USA

"Convergence is using the resources at a newspaper to bolster the product on other media. Convergence is using the newspaper story in other media. It is not a loss of voices. Convergence is making the best use of your newspaper. The whole objective is to have a fully informed public, so convergence is making news available in all forms to reach as many people as possible. You make the information work for people: One story serving more than one purpose."
> Keith Wheeler, associate managing editor for broadcast and online, Orlando Sentinel, Florida, USA

"Convergence in journalism means different things in different situations.... Convergence is generally seen in terms of increasing co-operation and collaboration between formerly distinct media newsrooms."
> Mark Deuze, University of Amsterdam, the Netherlands

The Emergence of Convergence

Figure 1. The CLTV news desk in the newsroom of The Chicago Tribune. Photo: Courtesy The Chicago Tribune

Rich Gordon of Northwestern University has identified at least five forms of convergence in the US. The first, ownership convergence, applied to the arrangements within one large media company that facilitated cross-promotion and content sharing among print, online, and television platforms owned by the same company. The News Center in Tampa, where the *Tampa Tribune,* WFLA-TV and TBO.com were all owned by Media General, represented probably the best-known example in the US. The Tribune Company, headquartered in Chicago, has long heralded the "synergy" that ownership convergence allowed. Gordon quoted its president, Jack Fuller: "Owning television, radio and newspapers in a single market is a way to lower costs, increase efficiencies and provide higher quality news in times of economic duress" (2003: 64). The state of the economy remains a key factor: Attitudes to convergence sometimes oscillate depending on whether the economy is doing well. Media economics are a direct reflection the state of the general economy. Tactical convergence was Gordon's term for the partnerships that had arisen among separately owned companies. Tactical convergence did not require common ownership.

The most common model was a partnership between a television station or cable channel and a newspaper owned by separate companies, and where each company retained its own revenues. It was a business arrangement in which each medium cross-promoted and marketed the other. "In most markets, the primary motivation for—and initial results of—these partnerships seemed to be promotional," Gordon wrote (2003: 65). An example was the relationship between *Florida Today*, based in Melbourne on Florida's east coast, and WKMG Channel 6 headquartered in Orlando, about 70 miles to the west. Ownership and tactical convergence have become common in the US. See the American Press Institute's convergence tracker for more details. Case studies and examples of convergence around the world are provided in chapter 5.

Structural convergence occurred when media companies reorganized the newsroom and introduced new positions. Sometimes this involved installing a multi-media desk in newspaper newsrooms. This form of convergence was associated with changes in newsgathering and distribution, Gordon wrote, but it was also a management process in the sense of introducing changes in work practices. An example was the *Orlando Sentinel*'s decision to employ a team of multi-media producers and editors to repackage print material for television. The team rewrote print content in a form suitable for television. They also produced original content (such as programs about the movies and high school sports), and arranged talkbacks between print reporters and the television partner. A talkback is a conversation between the television anchor and a reporter in the field or the office, where the anchor asks questions of a reporter who specializes in a particular beat. Keith Wheeler, the paper's associate managing editor for broadcast and online, described his job as getting "as much *Orlando Sentinel* content [as possible] to our media partners" (2004b).

Information-gathering convergence took place at the reporting level and was Gordon's term for situations where media companies required reporters to be multi-skilled (2003: 69). This represents probably the most controversial form: what some people have described as "platypus" or "Inspector Gadget" or "boat-car" journalism (Cole 2000; Dailey et al. 2003; Haiman 2001 and 2004). Most readers will be aware of the Inspector Gadget cartoon character, whose multitude of tools spring out of his hat. A platypus is a mammal found only in Australia: It has the body and hard, broad tail of a beaver, and the bill and webbed feet of a duck. It lives in a burrow near lakes and lays eggs, yet suckles its young. All of the above terms have been used in disparaging descriptions of the reporter who is required to do all forms of journalism but does none of them well.

A platypus scenario would look like this: After attending a news event a reporter would be expected to write a handful of paragraphs for the organization's website, produce a radio script, take still and video images, prepare a television package (or be available for interview as an expert for a talkback), and finally write a considered piece for the newspaper. Stories are sometimes also made available on mobile devices such as cell phones and personal data assistants (PDA), though in most cases software such as XML (eXtensible Markup Language) modified existing data in a form appropriate for PDAs and phones, and the reporter was not involved. The single multi-media reporter may be appropriate and workable at small news events. But at a major news event where scores of mono-media reporters outgun the single reporter, this form of reporting cannot produce quality work. Chapters 4 and 6 discuss this issue.

Bob Haiman, president emeritus of the Poynter Institute, said too many journalists and journalism professors seemed enthralled by convergence. "The more I hear, the less I am persuaded that it has much, if anything, to do with that obligation [to inform the public about the public's business and help to perpetuate self-government and democracy]." Haiman objected to the multi-skilled or "boat-car" form of journalism because it meant a journalist could not do one job well. The boat-car flopped, he said, because people discovered that while it was an ingenious combination of a car and a boat, "it was a lousy car because it also had to be a boat, and it was a lousy boat because it also had to be a car" (Haiman 2001 and 2004). By mid 2004 functional boat-cars were being sold in Florida. One should never underestimate the potential of technology to facilitate change. A more important debate centers on the quality of the content that journalists produce, and the potential or otherwise of convergence to serve democracy.

Digital technology makes the multi-skilled journalist possible, but in reality we may not see many Inspector Gadgets, for several reasons. Most journalists simply do not have the necessary level of technical expertise, and training has never been a high priority in many newsrooms. In terms of the quality of final product, it is not possible for one person to cover a major story adequately for all media. The "platypus" reporter may be appropriate for handling small or isolated news events or less timely stories such as features. If technology becomes significantly simpler, we may find more platypi emerging from their burrows, especially in circumstances such as war (embedded reporters in Iraq, for example) or isolated areas where it is not possible to send a team of reporters.

John Beeston, online news director for CNN Hong Kong, described how he sent one reporter to Kalimantan in the Indonesian jungle with a small digital video camera, mobile phone, and laptop in 2001. "She covered the story by telephone for CNN's international programs broadcast from Atlanta. In addition she reported into our regional programs that emanate from Hong Kong. She also wrote news stories and filed pictures for the web. When she returned to Hong Kong she brought back some sensational material, which we produced into enduring features." Beeston said small, lightweight equipment enabled the reporter to move around more easily than a crew of three people "with numerous metal boxes." His company was able to get elements of a story that would have been impossible for a traditional TV news crew. "This approach is not always suitable, but it gives us more flexibility" (Beeston 2001).

Gordon's final category was storytelling or presentation convergence. It operated at the level of the working journalist, though it needed management support in terms of purchasing the most appropriate equipment. Every new medium offers innovative ways to tell stories, but these conventions take time to evolve. Gordon predicted that new forms of storytelling would emerge from the combination of computers, portable newsgathering devices and the interactive potential of the web and television, as journalists learned to appreciate each medium's unique capabilities (2003: 70). Most reporters and editors are still working out how to "do" this form of convergence. Doug Feaver, executive editor of washingtonpost.com, said his journalists were "inventing a new medium" as they worked (Feaver 2004). One of the participants on the November 2002 Defining Convergence panel in South Carolina pointed out that convergence was still evolving and still an experiment: "If anyone is doing the work and considers it anything but research and development, they are way off" (Romaner 2002). Two years on, that situation had changed to a limited degree but convergence remains in the experimental or evolutionary phase in many newsrooms.

Gordon's five categories described convergence from the perspective of ownership and newsroom structures. A trio of academics from Ball State University noted that scholars and practitioners rarely defined convergence in terms of media behaviors. Larry Dailey said that an important question was not whether an operation was doing convergence, but the degree to which it was converged. He and his colleagues proposed a "convergence continuum" to provide a conceptual framework for understanding this innovation from the perspective of behavior. They identified five levels of activity among news organizations: cross-promotion, cloning, "co-opetition," content sharing, and full convergence. At the cross-promotional level the least amount of cooperation and interaction occurred among members of different news organizations; convergence was a marketing tool. At the cloning level, one partner republished the other partner's product with little editing. "News outlets at the cloning level do not discuss their news-gathering plans and share content only after a story has been completed" (2003: 6).

The term "co-opetition" is an amalgam of competition and cooperation and represents a new form of business in which organizations that originally were competitors work together when it suits both parties. Walter Keichel, editor of the *Harvard Business Review*, said the essence of the theory could be reduced to two sentences: "Cooperate with others to increase the size of the pie. Compete in cutting it up." But he inserted a cautionary note: "The others with whom you may wish to cooperate could include businesses with which you compete on other fronts" (Keichel 2001).

At the co-opetition stage, news organizations both cooperated and competed. According to Dailey, "At this level, the staff members of separate media outlets promote and share information about some stories on which they are working." Dailey noted that at this level, years of competition and cultural differences combined to create mutual distrust that limited the degree of cooperation and interaction. "For example, a newspaper reporter might appear as an expert or commentator on a television station's newscast to discuss a current issue, but the two staffs are careful not to divulge any information that might be exclusive to their news products."

At the fourth stage on the continuum, media organizations regularly published information gathered by the cross-media partner. Dailey and his colleagues called this content sharing. "The partners also might share news budgets or attend the other partner's planning sessions. Collaboration on a special, investigative or enterprise piece is possible. In general, however, the news organizations produce their own stories without helping each other." Full convergence took place at the right-hand end of the convergence continuum. At this stage the partners cooperated in gathering and disseminating news. "Their common goal is to use the strengths of the different media to tell the story in the most effective way. Under full convergence, hybrid teams of journalists from the partnering organizations work together to plan, report, and produce a story, deciding along the way which parts of the story are told most effectively in print, broadcast, and digital form." Teams gathered and produced content for specific projects and then disbanded, Dailey said, and new teams formed as projects developed (2003: 6–8). A continuum gave researchers the flexibility to identify and measure varying degrees of cooperation and interaction at news organizations, he said. The five circles on the continuum illustrated that a range of overlapping behaviors characterized each position as the degree of cooperation rose. "The arrows on the continuum show that a partner's place on the model is not fixed; it can move back and forth depending on the nature of the news and the commitment to convergence" (2003: 6).

For this author, convergence is more than cooperation (sharing of resources) or partnerships or cross promotion or content sharing—the main forms operating in the US. Full media convergence involves a radical change in approach and mindset among managers and journalists. It involves a shared assignment desk where the key people, the multi-media assignment editors, assess each news event on its merits and send the most appropriate staff to the story (see the new roles at the Newsplex discussed in chapter 4). Sometimes it will be an individual and occasionally that individual may even be an "Inspector Gadget" kind of journalist. But most of the time teams will be assigned.

Convergence coverage should be driven by the significance of the news event. That is, the importance of the story will dictate the coverage, and influence the size of the team involved and the depth and breadth of the reporting. Multi-media assignment editors will decide on the most appropriate media for telling the story. A major city fire may require a team of still photographers, video-journalists, online specialists, and reporters. A routine press conference may need only one reporter. Kerry Northrup, until August 2004 director of the Ifra-sponsored Newsplex (and the organizer of the South Carolina conference) noted that assignment editors—the people who allocated stories to reporters—were the key people in convergence journalism. Those editors needed a mindset free of any one medium: "A true multiple-media editor will be one who recognizes, for instance, that breaking news reporting is no longer a staple of printed journalism, and therefore that printed newspaper content must rise to a higher level while working in concert with its online siblings" (Northrup 2000b: 33). Northrup was adamant that convergence was not just cooperation. "Convergence of any consequence worth the effort is measured by whether it gives news consumers something more than was available to them before the media combined, by whether it results in some added value for the local news and information marketplace. If not, it is just an internal work flow exercise that will not significantly reposition the media company" (Northrup 2004b).

Juan Antonio Giner likes the analogy of the circus. He suggested that newspaper companies in the early twenty-first century were experiencing what happened to the circus business half a century earlier. "The one-man circus became a one-ring family circus and the one-ring family circus became the three-ring family circus," Giner said. But integration was merely "co-operation." Different family circuses, with different cultures (animals, clowns, musicians, singers, jugglers, and acrobats) shared the same tent, but in each ring they still were acting as a single circus. Real convergence only came when circuses mixed animals and people under the same tent and appointed a "three-ring master," he said. "My best advice is this: go to Ringling Brothers and Barnum & Bailey Circus and there, not at Florida newspapers, you will see the Greatest Convergence Show on Earth! They are the real integrators, not the U.S. newspapers that still are in the era of three separate rings" (Giner 2001c). As of the early twenty-first century, media groups in Scandinavia and Southeast Asia offered the best examples of three-ring convergence circuses. The World Company in Lawrence, Kansas, is the best example of convergence in the US, and is discussed in chapter 5.

Communication is a key factor in this full convergence process. If a story evolves to the point where one reporter is not enough, that reporter needs to be sufficiently flexible to know when and where to call for help, and sufficiently confident to know that the call for help will not lead to ridicule. People on the multi-media desk also need to be trained to assess a story and send the most appropriate individual or team. All information (image and text) must be fed into a central database from which relevant materials can be extracted to tell the story in the most appropriate way. Unused material must be archived to establish a knowledge base for future projects. Most reporters will remain specific to one medium, though the organization will employ a core of people expert in reporting in multi-media formats, and with time the proportion of these people will grow. These people will be prized. All budgets need to be linked so that each convergence partner knows what the others are doing and covering. Editorial managers need to know enough about the strengths of each medium to be able to discuss potential multi-media facets of stories.

All of this necessitates a change of mindset and attitude among media managers, and a willingness to learn more about the potential of each medium to tell stories in different ways. The team is more important than the lone wolf reporter because teams produce better multi-media reporting. It also means that news organizations will need to invest in widespread training, to teach journalists skilled in one medium how to tell stories in another, as well as spend money on equipment. This form of convergence is expensive and complicated. Larry Pryor of the Annenberg School for Communication at the University of Southern California noted that this was a huge task. It implied many things in terms of "cost/revenue sharing, news staff capable of functioning on all platforms, coordinated coverage, agreed-on deadlines, [and] shared newsroom space" (Pryor 2004). If we accept that people consume media for the quality and uniqueness of the content, which in turn gives those media the right to charge more for advertising, then media organizations are going to have to invest in the journalists who produce quality content. One of the big issues that must be resolved is whether media organizations are embracing convergence to produce better and more varied content, or as a way to market existing content to more audiences. It is appropriate to pause here to consider this potential dichotomy.

Business Models and Convergence

Viewed as a business model, convergence appears attractive to some editorial managers and publishers. At one extreme, they perceive that converged and multi-skilled journalists could potentially produce more news for the same or little more money. They then think that their organizations should be able to cut costs because of increased productivity: more multi-skilled reporting means the organization needs fewer reporters. This is one popular myth about convergence. Major opportunities do exist for cross promotion and marketing, where each medium recommends the next in the news cycle. But convergence does not cut costs in the content-producing areas. To illustrate the myth about saving money, Bob Haiman tells the story of asking an editor for the two words that best captured the essence of a converged news operation. "Without a moment's hesitation, eyes literally shining, he smiled broadly and said loudly: 'Better marketing'" (Haiman 2001 and 2004). But from the start we need to appreciate that convergence does not save money in the newsroom. James Gentry has advised several newsrooms on convergence, and written extensively on the subject. For him the bottom line can be summarized in one sentence: "Convergence always costs more than you think it will, takes longer than you think it will, and is more difficult to do than you think it will be" (Gentry 2004b).

From the journalist's perspective convergence offers a chance to do better journalism by giving reporters the tools to tell stories in the most appropriate medium. Technology frees them from the limits of individual media. Some print reporters are embracing convergence because appearing on television gives them added visibility—they enjoy being recognized in public places like supermarkets—and convergence skills also make them more marketable. They possess extra strings to their journalistic bows. Joe

Brown, a senior reporter at the *San Francisco Chronicle,* said convergence helped improve both forms of writing. "Across the board there is this misunderstanding between print and broadcast journalists. They sneer at each other. But once you've done it [broadcast journalism] you appreciate how difficult it is. You understand the limitations of broadcast [journalism] and the skills involved. You don't take it for granted and you also understand the limitations of the other medium. I think that convergence helps [print] reporters understand what's missing in their reporting. Print reporters who wind up doing television end up understanding what's missing in their work" (Brown 2004).

The potential dichotomy, then, is the conflict between a business view of convergence—multiple-platform publishing as a tool for increased productivity and marketing—versus journalists' aspirations where convergence offers them the potential to do better journalism. The latter form is unlikely to save money. One of the realities of American media is the need to make money. It is a fundamental truism of media economics that journalism needs advertising and advertising needs journalism: advertising pays for good reporting just as good reporting attracts customers for advertising. John Haile was the editor of the *Orlando Sentinel* who saw the potential of new media, and the dangers of media fragmentation. Haile emphasized the importance of protecting a company's revenues, and believed new media and convergence offered a way to do so. "The issue driving my actions was the threat to our ability to do great journalism. I had long believed that the soundest foundation of a free and successful press was a financially successful press. As I looked to the future, I could see how that financial base could be eroded by the proliferation of new media and the almost certain fragmentation of our audience" (2003: 4). Audiences were changing, with the consequent threat of declining revenues. "If ad[vertising] dollars start dropping, you can bet newsroom budgets will follow. That will dramatically affect our ability to do good journalism" (quoted in Gentry 1999: 6).

Haile had always been an innovator. In 1993 under his editorship the paper partnered with America Online to provide the local content and promotional arm for AOL in the area, and launched a website the next year. Also in 1994 the *Orlando Sentinel* formally established a new media operation. The paper also partnered with Time Warner to provide news, an entertainment guide and classified advertisements in the first test of interactive television, a project known as the Full Service Network. Haile introduced the concept of convergence in 1995 when a formal television partnership was added to the mix and a camera installed in the newsroom. "In the fall of 1998 we launched in partnership with Time Warner, the all-news cable channel Central Florida News 13 (CFN13). For this, reporters and editors contributed a regular stream of news stories throughout the day as well as feature packages." Later that year Haile designed and had installed a multi-media desk in the center of the newsroom (2004). Haile later became a partner with the Haile-Gentry consultancy group that advises companies on how to manage convergence.

In the best of all worlds the dichotomy mentioned above resolves itself in a balance where good journalism attracts enough advertising to sustain both the journalism and the need to make a profit. But when the equation is out of balance, such as during the economic recession of the early years of the twenty-first century, or when managers get greedy, the dichotomy is magnified. As long as the equation is balanced, convergence can work. How this dichotomy is resolved has profound implications for how all forms of journalism, including convergence, are practiced in the future.

Convergence and Digital Tools

Digital technology makes convergence possible, but these tools cost money and take time to learn. Kerry Northrup, from 2004 director of Ifra's publications division, points out that convergence does not save money: "The business side needs to understand convergence is not a cost-cutting strategy. It is a growth strategy." Convergence positioned media companies to expand their share of the market. "This is not to ignore that there are cost efficiencies to be realized through deploying editorial resources across a range of media activities rather than duplicating them for each media unit. However, it is unrealistic for a publisher to expect to go from working in one dominant medium to working in two, three or four without more people and technology." Businesses engaging in convergence to cut expenses would be disappointed by both the "financial and journalistic results," he said (2002a: 15). Convergence was an important new form of journalism, Northrup said, because it addressed the needs of the audience. "Convergence offers the audience new ways of absorbing news rather than just offering journalists new ways of presenting it" (Northrup 2000b: 32).

The convergence process also requires significant organizational change. Newspaper consultant Andreas Pfeiffer said newspapers were moving inevitably in the direction of convergence. But he issued a word of caution: "What has become clear, however, is that implementing this vision is far more challenging than may seem from a safe distance. Setting up systems for multi-channel publishing is a complex and costly task. While defining a multi-channel publishing system is relatively easy on the conceptual level, making it work is far more challenging" (Pfeiffer 2000). The Media Center at the American Press Institute in Reston, Virginia runs courses on convergence for both journalists and editorial managers. The center's director, Andrew Nachison, has identified the essential dichotomy and noted that the business approach was winning in the US: "I think journalism is adjusting to and coping with market forces and business imperatives—[but] I don't think journalism is leading the charge" (2002). In February 2004 Nachison refined his opinion, suggesting that media managers had become so caught up in the economics of the industry they did not recognize they were making decisions based on economic rather than journalistic principles. "So we hear editors talking about brand extension and market penetration. They have absorbed the business jargon. The industry has co-opted business models and approaches. This is a more troubling trend, though in some respects it's nothing new" (2004).

Convergence is attractive to both media managers and practitioners because it satisfies consumer demands and lifestyles. It also protects an organization's journalistic franchise in the sense that multiple-platform publishing—increasingly an alternative phrase for the more nebulous term "convergence"—allows wider coverage of an area and permits cross-marketing of a single product. Organizations embrace multiple-platform publishing for a variety of reasons, and produce multiple forms of convergence. It would be safe to say that no two media groups produce the same form of convergence. Their reasons are connected with the values and background of the people making the decisions. The influence of values in terms of how people define convergence is discussed in the next chapter. Managers see the financial benefits of having staff expand their expertise in many formats; meanwhile journalists in those organizations believe quality must be maintained and call for this to remain the paramount consideration in any change (Quinn 2002: 59).

Consultant and former editor Katharine Fulton concluded that smart journalists would embrace new forms of journalism such as multi-media storytelling and find ways to employ their traditional abilities to synthesize, explain and place events in context. "They'll also work to re-interpret those old values for a new era" (Fulton 2000: 35). Journalists in the twenty-first century will need a flexible mindset and the ability to adjust to change. This was the conclusion of Brian Veseling, deputy editor of the respected industry magazine *newspaper techniques:* "If there is one word to describe what is being required more and more in newsrooms as journalism moves into the digital age it is flexibility. In an industry in which flexibility always has been an important element for success, it now seems to be vital for survival" (Veseling 2000: 20). Convergence is about both survival and change.

Forces Driving Convergence

Several forces are driving convergence, and the next chapter discusses them in detail. The key forces are the changing lifestyles of news consumers and consumers in general, and the rise in the amount of time people spend with various forms of the media. As the *Tampa Tribune*'s Thelen has pointed out, people's information-seeking behaviors are changing and media organizations need to adapt to respond to that need. Howard Tyner, a former editor of *The Chicago Tribune* who became a senior vice president of the Tribune Company, has maintained that the business of journalism was about "eyeballs"—getting as many people as possible to look at media products. A media company's aim was to deliver content to consumers. "We need for our news and information to go to the eyeballs of web consumers and TV viewers and cable customers and even radio listeners although they aren't using their eyes. We go where the audience is," he said (Tyner 2004). In essence, convergence increases an organization's chances of reaching the largest number of eyeballs.

The change of news habits needs to be understood in the context of increasing consumption of media in the Western world, combined with shrinking markets for individual media. The 2004 *Communications Industry Forecast* showed that in the US in 2003 the average consumer spent 3,663 hours a year using all forms of media. That's more than 10 hours a day reading, listening, watching and surfing for any combination of professional and personal reasons. It represented an increase of almost an hour a day since 1998, the report said (P. T. Larsen 2004). Ifra's Northrup said the American situation could be transposed to other developed economies. "It is one of the reasons the U.S. Federal Reserve, the European Central Bank and most other monitors of the world's financial pulse say we have now officially transitioned from the industrial economy to an information economy" (2003a: 3).

As well as using a lot more media, people are using it in multiple forms to fit everything into their busy days. James Rutherford, executive vice president of the company that published the 2004 *Communications*

Industry Forecast, pointed out that consumers were using two or more media simultaneously to cope with the range of media choices and the competition for attention. "The result is a media generation consuming more information in less time than ever. Time is the most precious commodity." Analysts at Rutherford's company, Veronis Suhler Stevenson (VSS), predicted the time an average American spent with media would increase by another hour a day by 2008. The previous VSS forecast published in 2003 said that by 2007 American consumers would devote 10.6 hours a day to the media, more than any other activity in a given day (quoted in Northrup 2003a: 3).

In October 2003 BIGresearch of Columbus, Ohio, published a study showing that almost three in four consumers used multiple media simultaneously. The study found that 74.2 percent of people regularly or occasionally watched TV and read the newspaper at the same time, and 66.2 percent regularly or occasionally watched TV while going online (API 2003). Research from a group of Ball State University academics revealed that people in Delaware county, in which the university was based, consumed 10 hours of media a day, and at least a quarter of that time was spent absorbing more than one medium at a time. Given the nature of the studied area (famously known as Middletown because of the depth and breadth of research conducted in the community since the 1920s), the researchers believed it was possible to extrapolate their findings to the national level (Papper, et al. 2004: 25).

At the same time that media consumption was growing, individual news media were seeing their market share decline, and people were more willing to pay for media. The 2003 *Communications Industry Forecast* reported that the media that consumers subscribed to (such as paid-content television) continued to grab market share from advertising-supported media (such as newspapers or network television). The former gained 10 points between 1997 and 2002 (VSS 2003). Concluded Ifra's Kerry Northrup: "More people are consuming more news and information content than ever before. And they even seem increasingly inclined to pay for it. Forward-thinking news organizations are responding to this opportunity with strategies that involve combining content formats and delivery channels to tap into the new audiences and to better mirror the way contemporary news consumers integrate print, video, online and mobile to satisfy their need to be informed" (2003a: 3).

Ruth de Aquino, Ifra's director of editorial strategy until 2002 who became a newspaper manager in Brazil, noted that the public's consumption of news had changed dramatically in the early twenty-first century, compared with the early 1990s. "News information is all around: on mobile phones, newspapers, portable data assistants such as Palm Pilots, television, interactive and cable TV, the Internet, teletext, kiosks [units which display the news in public places], radio, video screens in hotel elevators, video programming for airlines and much more. The concept of news is changing all the time. [It is becoming] more personalized, more service-oriented and less institutional" (2002: 3). In a fragmented market the ability to reach as many potential audiences as possible becomes highly attractive, and may ultimately decide whether a news organization survives. As examples later in this book show, convergence makes it possible to reach more audiences. The advertising pie—the total amount of money available to media organizations—may not change much. Sometimes recessions may even reduce the size of the pie. Convergence improves a media company's chance of reaching as many people as possible. Advances in digital technology make convergence more possible technically, and these advances will accelerate as broadband becomes more widespread. The next chapter discusses them in detail.

Fostering Convergence

It is possible to tease out several factors common to the most successful converged media operations. The next chapter discusses them in detail, and the final chapter offers a way to implement convergence within the newsroom and the boardroom. This section summarizes them for people eager to move on to other parts of the book. The most vital factor is management buy-in, in the sense that management is seen to support and expect convergence. At the *Orlando Sentinel* in Florida, for example, new recruits join the paper with the understanding that they will operate as multi-skilled reporters. The core of the converged newsroom is information, and an appreciation that modern reporting should not be bound to traditional notions of journalism as a linear process devoted to one medium. Rolf Lie, editor of Norway's *Aftenposten*, believes the future is not about paper or electronics, but about information. "Today's journalist should say: 'I'm not working in a newspaper, I'm working in news,'" he said (Lie 2000: 1). This involves a change of mindset, which is another of the factors common at news organizations that have successfully embraced convergence. It is a manager's responsibility to foster that mindset.

In Hong Kong the *Ming Pao*'s chief editor Paul Cheung said the move to a multi-media environment could only be successful if accompanied by a corresponding change in the attitudes of journalists. The transformation must occur first in the minds of editorial managers: "From my point of view, the chief editor has an important role: He must be a leader in terms of the changes" (2001). Ulrik Haagerup, editor of *NordJyske* in northern Denmark, looked at the situation from a different perspective but came to the same conclusion about mindset. "It is only in English that 'newspaper' has something to do with paper. Media convergence has nothing to do with technology or architecture. It has everything to do with mindset. People out there are moving fast. They are changing the way they use their news media and we have to change with them. Put the customer first," he said (2002b).

Another common factor involves placing people with different skills in the same physical space to generate trust and the sharing of ideas, which leads to synergy. At *Ming Pao* chief editor Paul Cheung said that parent company Ming Pao Enterprise Corporation hired senior journalists and photographers to help with the move to multiple-journalism (Cheung 2001). News organizations need to flow information and content through the organization in such a way as to make content available for multiple platforms. Forrest Carr, news director for WFLA-TV in Florida, which partners with the *Tampa Tribune* to provide converged news, said all the Tampa newsrooms shared story ideas via custom-built software called Budget Bank. (Late in 2004 the company was trialing a new browser-based platform integration product from CCI-Europe called NewsGate). Intranets also permit the easy distribution of information (Carr 2004). These and other web-based technologies are vital because data and information are the lifeblood of modern media organizations.

Cultural factors can encourage or inhibit convergence. A conservative newspaper that sees itself as a paper of record will have difficulty partnering with a tabloid television organization. Saints will not work with people they perceive as Satans. Similarly print reporters who look down on television journalists, perhaps mocking them as poor spellers who only skim the surface of news, are less likely to welcome broadcast people into their newsroom. Ifra's Northrup said that cultural issues, the stereotypes that journalists held about other media, and developing an appropriate mindset among journalists were the most difficult but essential concerns that managers needed to deal with when converging newsrooms and staff from competing media (2004b). The Poynter Institute's Howard Finberg noted that no single culture existed in a media organization. "There are multiple cultures throughout a news organization. And when it comes to convergence, you can add several more groups, such as broadcasting with its on-air talent, production staff, and producers."

Part of the problem is the common language that separates print and broadcast journalists. The glossary at the end of this book attempts to help with this issue. An editor at a newspaper, for example, performs a very different role from an editor in a television newsroom, and redundancy in a newspaper newsroom has a very different meaning from redundancy as discussed by technology writers. Finberg said convergence increased the complexity of cultural relationships "within a newsroom, within a company, and within the corporate boardroom" (2002). Part of the answer involved exposure of shared values, where journalists learned to trust people from other media by working with them and discovering common values. The manager's role in this situation is to point out to reporters in different media that they share the same core values. Another key is training, in the sense of exposure to ideas and learning how to operate in different media.

The Key Role of Managers

The importance of quality managers who support and promote convergence cannot be overemphasized. Those managers must appreciate that technology is a tool for doing better journalism, and journalists need training to use their tools more effectively. Encouragement must come from the top. Editorial managers must be willing to foster an environment that facilitates life-long learning. One of the key roles of journalism in the forthcoming knowledge age will be to turn information into knowledge—to synthesize it for their audiences. This takes training, which requires an investment of time and money. Respected MIT economist Lester Thurow has condemned employers in the US for their attitude to training. "The basic problem in the United States," said Thurow, "is that every employer wants to free-ride the training system." Whenever unemployment levels fell, he said, companies complained about the shortage of skilled workers—yet these complaints came from companies that did not train. "They know that they need a better trained workforce but think that someone else should take the responsibility for, and bear the cost of, creating it" (1999: 269). Managers in Denmark have a different attitude to training. Danish journalists are entitled as part of their union award to a week of training a year, and many use it to work toward a degree or other qualification. The final chapter will discuss the importance of training if convergence is to happen, and chapter 7 looks at the role of knowledge management in helping to create quality content through development of the smart newsroom.

It is important to appreciate that innovation takes time and often generates resistance. Paul Saffo, director of the Institute for the Future in Menlo Park, California, applied what he called the 30-year rule to change. For the past 500 years, he said, it had taken about three decades (about a generation) for a new idea to "fully seep into a culture" (1992: 18). New media analyst Roger Fidler wrote that slowness of change was a "rule rather than the exception" with emerging technologies. "The 30-year rule may not be foolproof, but it does put the development of new technologies into a more realistic perspective" (Fidler 1997: 10). Convergence is already happening in many newsrooms, especially in Scandinavia, South America, and Southeast and South Asia (see chapter 5 for case studies). For organizations wanting to introduce convergence, the final chapter of this book offers a process for implementing it based on lessons learned from around the world.

Convergence produces many challenges for journalists and publishers. Both groups need to find a way to gather and fund news for different platforms without compromising the needs of their audiences, and while maintaining ethical business practices. In essence they need to find ways to blend the twin aims of telling the truth and making money. Ari Valjakka, editor-in-chief of one of the world's most successfully converged media companies, the Turnun Sanomat Group in Finland, has warned of the danger of trivialization of quality journalism when a journalist "shovels" a story from one medium to another without producing content appropriate for the medium. "But when you utilize the strengths of different media—speed of [the] web, text-TV and radio; 'visuality' in television and background material in print—this danger turns into strengths" (2001).

Managers' editorial and social values remain a key issue. The role of journalism is to help citizens make choices in an increasingly complex world. Or as Kovach and Rosenstiel put it so elegantly in *The Elements of Journalism*, journalism is a process of "sense-making based on synthesis, verification and fierce independence" (2001: 197). If managers introduce convergence as a way to save money, reporters may become too busy to verify the information they find and resort to publishing material supplied by professional "spin doctors." Journalists could become so busy providing content for multiple platforms that they simply do not have the time to reflect or analyze. But under wise and ethical leadership, convergence offers opportunities to do better and more socially useful journalism. The next chapter looks in detail at the forces bringing about this revolution in journalism.

2 Why and How Convergence Is Emerging

The first half of this chapter looks at the bigger picture and describes the social, technological, and legal issues producing change in society and the media that reflects it. The second half considers how convergence happens, and describes the circumstances that lead to its successful introduction to the newsroom. This section also discusses the barriers to convergence, highlighting issues that have made introduction of convergence more problematic. In essence, this chapter will look at:

* Social factors affecting convergence
* Technological factors affecting convergence
* Legal and structural factors affecting convergence
* Macro to micro: Factors that make convergence possible

It is important to understand the environment in which the seeds of convergence are beginning to grow. This chapter begins by looking at the major changes in society as the twenty-first century emerges, and the impact those changes are having on the business and practice of journalism.

Social Factors Affecting Convergence

Many people in advanced societies in the early twenty-first century tend to be time-poor but asset-rich. One consequence of this evolution is the tendency to expect and need convenience. People usually are willing to pay for it—witness the boom in online commerce, drive-through services, the wide range of take-out food outlets, subscription services, and the boom in purchase of labor-saving devices. In some parts of the US the battle for people's disposable time has become more vigorous than the fight for their disposable income, noted *Wall Street Journal* consumer reporter Martin Peers. In some key demographic groups, time has become more scarce than money. In the last 30 years of the twentieth century (about a generation), the median number of hours that American people said they worked a week jumped from 41 to 49. Harris Interactive, which each year surveys adults about their work and leisure time, said the increase had been at the expense of people's leisure, which dropped from 26 to 19 hours a week over the same period (Peers 2004: D1). In 2003 employed Australians averaged 1,855 work hours a year, narrowly edging out American and Japanese workers (1,835 and 1,821 respectively) for the title of the people who spent most time at work (Tiffen and Gittins 2004: 82).

NBC's head of research, Alan Wurtzel, has suggested that the years between 2003 and 2005 will be regarded as a "watershed" in social life, and the "beginning of a very different era" in terms of attitudes to time. A generation ago families watched television together, but by the start of the twenty-first century that concept had become so unusual that television executives gave it a special term, "co-viewing." In 2003 the Pew Center reported that people were watching less television and spending more time on the Internet. Mark Hinojosa, assistant managing editor for electronic news at *The Chicago Tribune*, said people found it difficult to carve out the time needed to appreciate a newspaper the size of the *Tribune*. "We know there are people out there who want their news from the Internet. I think the battle [for people's time] that newspapers can win will be weekends, and the battle they could lose will be during the week" (Hinojosa 2004). Products and technologies that people can use while on the move, such as the cell phone, or which allow consumers to manage and maximize their time, such as digital video recorders (TiVO is the best known brand), have become increasingly popular. "Multi-tasking is clearly spurring sales of Apple Computer's popular iPods and other portable MP3 players, and no doubt explains the rise in hours spent listening to the radio and using wireless phones" (Peers 2004: D1).

At the same time, people have to deal with information overload, vast amounts of data and information flooding into their lives. But this may not produce any more understanding. In the poem "The Rock" T. S. Eliot asks: "Where is the knowledge we have lost in information?" The amount of information being generated each year is so huge that it is difficult to appreciate. The University of California at Berkeley's School of Information Management and Systems calculated that in 1999 the world produced between one and two exabytes of unique information, or 250 gigabytes of unique information for every man, woman, and child on the planet. A gigabyte is about equal to a truckload of documents. This represents about 250 truckloads of printed pages for each person, and the volume of data and information is increasing. Each year the world produces another 250 of these truckloads of new information for each person, noted writer Philipp Harper. "It is no exaggeration to say that information, so omnipresent and pervasive—so available—has been reduced to the level of white noise, a persistent buzz always in the background of modern life" (Harper 2004). In *Data Smog: Surviving the Information Glut,* David Shenk argued that the "paramount challenge" for the modern journalist was to learn to share and cultivate information, and to make it manageable for audiences. "This is not so much fact hunting as it is data gardening," he said (1997: 168). A world of excess information necessitated a restructured value system in which sharing and summarizing existing information was more of a priority than stumbling onto genuinely new data. The traditional news paradigm had become outmoded, he said. "New information for its own sake is no longer a goal worthy of our best reporters, our best analysts, our best minds. Journalists will need to take a more holistic approach to information as a natural resource that has to be managed [rather] than acquired" (1997: 170).

Changing News Consumption Patterns

One consequence of the drive for convenience is a change in the way people get their news, and the way they interact with it. Newspaper manager Ruth de Aquino touched on Shenk's idea when she suggested that the concept of news was changing and becoming "more personalized, more service-oriented and less institutional" in a presentation to the World Association of Newspapers (2002: 3). Jack Fuller, president of the Tribune Company, noted that people used to buy papers such as *The Chicago Tribune* as a "seven day a week habit." By the start of the twenty-first century many consumers had got out of the habit. Fuller said people bought papers such as the *Tribune* when "something happens in the world." The rest of the time, he said, they tended to look for more convenient options such as the Internet. Fuller suggested that providing free information on the Internet could have been one of the newspaper industry's biggest mistakes: "It's easy to persuade people that information is free," he told the November 2003 annual conference of the Online News Association in Chicago. But he noted the strategy was necessary to build audiences. "For rational reasons we got everybody used to the idea that high quality news ought to be available for absolutely nothing."

For a variety of reasons, consumers have come to expect free and convenient information on the Internet that is available in an easily-digestible form. Audiences were attracted to short and immediate pieces of information that had punch, Fuller said. It was sobering to note the difference between what his staff put on the front page of *The Chicago Tribune*, and what Internet access logs showed people went to the Internet for. "We can reach young people with what we like to think of as news. We can reach them so long as we don't make the reports too demanding or long or difficult." *RedEye*, the Tribune Company's daily paper that targeted the youth demographic, was the product of what the company had learned from interactive experiments, Fuller said. "People actually wanted a fair dose of hard news. They also wanted entertainment. The online medium does not lend itself to long-form discourse. Our research suggests there is almost no audience for that kind of journalism [like putting an extended document such as a president's full speech on the web]." The result is a society accustomed to convenient news, that is, news "online, any time and when they want it" (Fuller 2003). In the middle of 2004, *The Sunday Missourian,* one of the newspapers produced by the prestigious journalism course at the University of Missouri, was experimenting with new formats. Stories were either very long or very short, based on the principle that people wanted a quick fix of news, or they were willing to invest their time in something that interested them.

Audience Fragmentation

Media audiences are fragmenting as people choose to grab their news when and where it is convenient. Take the example of executives aged 45 to 54, traditionally the people who get much of their news from magazines and newspapers. *The Wall Street Journal* reported that almost all senior-level executives in a survey of Fortune 1000 companies used the Internet extensively at work and home, and subsequently spent less time with every other medium. The survey, conducted by Harris Interactive, revealed that 99 percent of senior executives accessed the Internet at work, while at-home use jumped nearly 10 percent to 97 percent in 2004 compared with 1997. The average age of respondents was 50. Mike Henry, director of sales for the online edition of the *Journal*, which commissioned the survey, said newspapers remained executives' primary news source, but they tended to read them for analysis rather than breaking news, for which they relied on the Internet. A quarter of respondents described the Internet as their most important news source. The biggest increase in time spent online, Henry said, occurred among the group aged 55 and older (Kaye 2004: C1).

Journalism was also experiencing big changes. One of the authors of the year-long study of the state of journalism in the US released in 2004 said the profession was going through changes as huge as those produced by the invention of the telegraph or television. The study was funded by the Pew Charitable Trust and conducted by the Project for Excellence in Journalism, an institute affiliated with Columbia University graduate school of journalism. Dante Chinni, one of the study's authors, said the lines between cable television, free-to-air television, and print media were beginning to blur. At the same time media audiences were fragmenting. Network television was moving more and more online. Newspapers were already online and the Internet was almost ubiquitous. "When the media fragment too much, our ideas of what is reality [also] fragment," Chinni said, noting that the printed media had doubled the amount of entertainment and celebrity gossip as a survival strategy. Matthew Felling, media director of the Center for Media and Public Affairs, said news forms were also changing. News was losing its hard edge and becoming what Felling called "cotton candy disguised as news." He questioned whether news organizations were looking to widen their audiences or merely attempting to attract younger viewers (quoted in Landphair 2004).

Newspaper consultant Juan Antonio Giner questioned the intelligence of offering simplistic, entertainment-based news, pointing out that newspapers were in the "credibility business." Convergence was an opportunity to return to basics, but with one important difference. "Back to the basics no longer means back to the old practices." It was an opportunity to reinvent journalism through convergence. The "information engine" or the "24-hour newsroom" was the new starting point of this recovery, he said. Traditional values had to prevail: "Only . . . by preserving the 'journalistic soul' of our newspapers will it be possible for newsrooms and management to work together on integration that is vital for the future, and that understands that, for the profitability of the news industry, the central pillar is the newsroom" (Giner 2001a).

Gil Thelen, publisher of the *Tampa Tribune*, agreed that convergence focused on the newsroom. "The quality of the journalism on all three platforms [at the News Center in Tampa] is better than what it was four years ago [in 2000], and the work is more sophisticated. Nor is it a fusion of brands. They remain separate and I do not foresee changes." Thelen is committed to the values of journalism, and has worked hard to ensure that the quality of his newspaper's reporting remains high in the move to multi-platform publishing. Poynter Institute scholar Howard Finberg pointed out that convergence was a reflection of society's move into the information age. "Convergence is a process: It is a journalistic process

[and] it is a business process." Before convergence could succeed in the newsroom, Finberg said, it had to be adopted in the boardroom, a place where major cultural and business changes were needed. "That's a pretty important role to assign convergence" (Finberg 2003).

Convergence offers a way to satisfy the audience's desire for news 24 hours a day, seven days a week, in multiple formats to reach multiple audiences. The aim is to supply news and information whenever and wherever the audience wants it, in as many possible forms to reach the maximum number of people. Paul Horrocks, editor-in-chief of *The Manchester Evening News* in the UK, summarized the situation in which journalists operated early in the twenty-first century: "If to information overload we add the fragmentation of media consumption, the drop of advertising revenues and the increase in competition then the conclusion is obvious. We must reinvent our product to be more competitive and satisfy our consumers." Convergence was the answer, he said, because it reached audiences eager to receive information in a variety of forms (quoted in Pascual 2003a: 34).

John Sturm, president of the Newspaper Association of America, told the group's 2004 meeting that traditional "ink on paper" companies had a chance to transform themselves into more broad-based media, given that an economic recovery was taking hold. Sturm said people were demanding instant content on what mattered to them and newspapers had to exploit technology to provide that content. "People want to consume their media where, how and when they choose," he said. "They want to look only at the ad[vertisement]s that interest them. Until someone comes up with TiVo for newspapers . . . we are the quintessential pull media." TiVo is a box attached to television sets that records programs on a hard disk, letting viewers "pull" what they want instead of networks "pushing" programming in the usual way (quoted in McDonough 2004).

Technological Factors Affecting Convergence

At the same time that audiences are fragmenting and demanding news when it is convenient, a major change is occurring in terms of the technology that brings information into homes and offices. In many parts of the world people have accepted online as another media form. In the US, access to the web had reached significant levels as of early 2004, with almost three in four people reporting Internet access at home and the vast bulk of businesses providing fast access at work. A Nielsen/NetRatings survey collected in February 2004 and reported the next month noted that about 204.3 million people, or 74.9 percent of the population, got Internet access via telephone modems at home. This was a big jump from the 66 percent reported a year earlier.

Kenneth Cassar, director of strategic analysis at Nielsen/NetRatings, said that in a handful of years online access had managed to gain the kind of "traction" that it took other media decades to achieve. Four in five people aged 35 to 54 had access to the Internet, closely followed by the 25-to-34 age group, where three in four people (77 percent of women and 75.6 percent of men) could go online. The demographics suggested that online had already become an accepted way of accessing media, and would continue to be so. "Women make the majority of purchases and household decisions, so it's no surprise that they are utilizing the Internet as a tool for daily living," Cassar said (quoted in Reuters 2004).

A boom in Internet use is likely to occur when broadband becomes more common. Broadband users can quickly download bandwidth-greedy content such as music, video and other multi-media offerings, which tend to choke traditional dial-up lines. Broadband will boost access to rich content along with acceptance of convergence. The availability of broadband continues to rise in many parts of the world. In South Korea, the country where broadband was the most widespread in the world, somewhere between 71 and 73 percent of households subscribed to high-speed Internet via broadband as of mid-2004. Most of the connections were in the big cities because of the concentrated nature of the population. News services such as OhMyNews.com have revolutionized the way people used media in that country (see chapter 8 for details). Nations like South Korea and Japan have made deployment of broadband services a national priority.

Reports vary as to the extent of broadband connections in the US. In April 2004 the Pew Internet and American Life Project reported that 55 percent of all adult Internet users—about one in three of adult Americans—had access to high-speed Internet connections either at home or work. Two in four adult Internet users—or almost a quarter of all adult Americans—had high-speed access at home, an increase of 60 percent since March 2003. October 2004, broadband adoption at home had moved past dial-up use in the United States, reaching 53 percent of residential consumers (Pew 2004). Point Topic, a London-based broadband research company, reported that nine in 10 American households had the potential to receive broadband though only two in 10 actually subscribed (Rosenbush 2004).

Regardless of which report we accept, trends show that increasing numbers of people are subscribing to broadband. In 2003 the number of American homes with high-speed Internet service jumped somewhere between 4.3 million and 8.3 million, depending on source data. The total reached somewhere between 23.1 million and 28.2 million households by the end of that year, again depending on sources. The more conservative figure was from Forrester Research, while the Federal Communications Commission (FCC) reported the higher figure. Cable-based services were the most popular form of broadband in 2003, the FCC reported, representing 16.4 million lines, or 66 percent. Digital subscriber line (DSL) accounted for 9.5 million lines, while the remaining 2.3 million high-speed lines used satellite, fiber-optic or wireless technologies.

As part of a plan to give all Americans high-speed Internet access by 2007, the FCC is looking at ways to establish wireless as a viable alternative to cable and DSL. The FCC was considering the reallocation of spectrum from broadcast television to wireless and expanding bands in the 5GHz range, FCC chairman Michael Powell told the Wireless Broadband Forum in Washington in April 2004. Wireless services were expected to improve that growth significantly. Powell said wireless would play a "critical" role in broadband developments, noting that it could reach consumers in ways that wired services could not. Providing a variety of technologies would decrease the potential for monopoly control and help avoid technical bottlenecks. "All the raw material is there, the recognition is there, and the understanding of its importance has begun to gel," Powell said. "Now, all that is left is the easy part of actually making it happen" (quoted in Shim 2004).

Factors likely to boost the uptake of broadband include the attraction of faster connection speeds, falling prices, and the bundling of services (for example, by combining video, data, and telephone services). "Cable's combination of higher access speeds, wider availability and more compelling bundles lets the leading operators target several classes of consumers," noted James Penhune, director of Strategy Analytics' broadband media and communications practice. "These include new users as well as a growing number of consumers seeking advanced video services such as high definition television and video on demand. At the same time, aggressive cable telephony deployments from leaders like Time Warner, Comcast, and Cox will help these companies reach customers seeking simplicity and savings from triple-play bundles combining video, data and telephony" (Penhune 2004).

Forrester Research analyst Ted Schadler wrote the company's Consumer Technographics 2004 North American benchmark study, which involved surveys of 60,000 households. Schadler noted that broadband subscription costs declined from $60 a month to as low as $25 in 2003. The demographic makeup of broadband users was attractive to advertisers, he said. People in broadband households earned 27 percent more than dial-up users, were online 52 percent more of the time, and spent more when shopping online ($80 a month more than dial-up users), he said. Commenting on the report, senior analyst David Hallerman for eMarketer said broadband was becoming more attractive to advertisers. "The speed is a large part of it, but the always-on nature is just as important," he said. "The Internet becomes a much more integral part of the overall media experience" (quoted in Dobrow 2004). Broadband adoption is discussed in more detail in chapter 8.

Cheaper Newsgathering Tools

The relatively low cost of portable newsgathering tools is similarly changing media managers' attitudes to the gathering of content. For years, commentators have suggested that journalists would never compromise on quality. They maintained, for example, that newspapers would never use inferior images generated by cell phone cameras, or that low-grade video would never be used on broadcast television. But news values tend to be more important than production values if an event is sufficiently newsworthy. On 18 February 2004, *The New York Times* published a photograph on its front page taken with a cell phone camera. It was an image of AT&T CEO John Zeglis signing the Cingular-AT&T Wireless merger agreement in New York the previous day. Joseph McCabe Jr., AT&T's chief financial officer, snapped the event with his cell phone's camera. The merger signing took place just before 3 A.M., and no photojournalists were present. One could argue whether the signing of a merger represented a news event worthy of page one, but it was a milestone in the use of a convenient form of technology for newsgathering. It was the first time *The New York Times* used a cell phone photograph on the front page (*The New York Times* 2004: 1).

In November 2003, the British Broadcasting Corporation (BBC) also changed some long-standing practices because of the availability of simplified newsgathering technology. It gave 40 of its reporters and producers cell phones that could record and send video. The BBC asked Philips Software, a division of Royal Philips Electronics, to reconfigure the Nokia 3650 cell phone so that its picture resolution was good enough to broadcast. The phones were able to record several minutes of video, instead of the 10 to 30 seconds available on consumer models. Justen Dyche, who coordinated the launch of this technology, said the phones were different from the satellite phones that networks had used to send live broadcasts from Iraq and Afghanistan. The latter delivered pictures taken by standard video cameras via satellite, while video cell phones sent their images via the cellular network. The BBC's head of newsgathering, Adrian Van Klaveren, said the phones were not intended to replace traditional television cameras but to augment them (quoted in Patsuris 2004).

In November 2003, BBC journalists became the first in the world to employ innovative software to broadcast video news live via laptop computers. Laptop newsgathering (LNG) requires a digital video camera, a laptop, and some proprietary software called QuickLink. The software needs a high-speed Internet connection such as a satellite phone, ISDN line, or wi-fi connection, and software delivers footage and reports using Internet protocol (IP). Peter Mayne, executive editor of BBC newsgathering, said the BBC had provided Quicklink to all its reporters around the world. "The system was used extensively during the Iraq war by our news teams who were in the most forward positions. Being in the thick of the action [they] needed to travel with the smallest and lightest equipment possible." Mayne said the BBC could easily update the software to its correspondents. It was scaleable depending on the available Internet connection, and could operate from about 64 kilo-bits a second through to one mega-bit a second. "The greater the bandwidth, the better the picture quality," Mayne said (Quicklink press release 2004). Ken Herron, director of Quicklink, said several other broadcasters were experimenting with the software. "Journalists working anywhere in the world could film a story, upload it to the web and then try to sell it to broadcasters," he said. Videophone connections could only transmit live video at a maximum speed of 128 kilo-bits a second, Herron said, while Quicklink software allowed feeds of up to one mega-bit a second depending on the speed of the connection (Kiss 2003c).

The BBC is also embracing the concept of the "video journalist," a single person in the field who reports, shoots, writes, edits, and transmits stories. The program, known as PDP or Personal Digital Production, began in September 2001. The BBC established a training center at its BBC Newcastle office and planned to train about 600 of its more than 5,000 journalists. As of early September 2004, 475 staff had been trained.

Each course lasted three weeks. Training was scheduled to end in March 2005. Michael Rosenblum, a former NBC producer turned consultant, developed the idea at NY1 in Manhattan and convinced the BBC that it could boost newsgathering efficiency by using video journalists. Rosenblum said the scheme was an attempt to "build television along the

Figure 2. The cluttered and busy newsroom of BBC Interactive in London. Photo: Stephen Quinn.

lines of a newspaper" operation. "We want to take them out of the newsroom and put them in the field where they can gather news," he said. Rosenblum said his process cut the cost of production by 20 to 70 percent (quoted in *Broadcast Engineering* 2004).

Paul Myles, the PDP center coordinator based at BBC Newcastle, said video journalists mostly used Avid DV Express 3.5.4 nonlinear editing software, though some offices operated Macintosh-based systems and edited with Apple Computer's Final Cut Pro. Three offices equipped with PCs used Liquid Edition. Video journalists were initially given the Sony PD150 digital video camera. From 2004 course attendees received a later model, the PD170. "It's a lightweight camera that records in DVCAM and has two channels of audio. We make several alterations to the basic camera. We have replaced the onboard Sony domestic microphone with a Seinhesser 416 microphone. It's a sensitive and directional microphone that helps us acquire excellent actuality." Myles's team also added a wide-angle lens and lens hood. "This allows us to get closer to the subjects we are filming, providing the benefits of a steadier shot, better depth of field, clearer audio and greater intimacy with character."

Myles said video-journalists mainly contributed to the BBC regional evening news programs but they also filed to current affairs, political, Welsh language, and children's programs. "The range of stories and techniques are almost as numerous as the trainees themselves. Many find the access and the ability to tell stories through real people's eyes the big attraction. For the others, multiple deployments are a big draw offering the ability to show several dimensions of a story simultaneously." Myles said the flexibility offered by the nonlinear editing systems helped producers create "very individual styles." Video journalists were not intended to replace television news crews, but to supplement traditional ways of working and to offer more "up-close-and-personal" stories. "It is inevitable that the use of 'self operating' staff will reduce the use of traditional crews but this wasn't the reason for doing it. The big attraction was that this way of working would give greater access, more freedom and creativity to the video-journalist, and a more honest and interesting final product" (Myles 2004).

Legal and Structural Factors Affecting Convergence

Technological change invariably runs well ahead of changes to laws because regulations take time to be implemented, while technology is always changing and moving forward. But regulation is a key factor in the emergence of convergence, in the sense of providing a framework for its evolution. Singapore provides an example of the influence of legal factors in determining the development of convergence. Until late 2000, the country's two media giants operated a comfortable duopoly: Singapore Press Holdings (SPH) ran all print media and the Media Corporation of Singapore (MCS) provided all the broadcasting. SPH published eight dailies: three in English, three in Chinese, one in Malay, and one in Tamil. In 2000 SPH started two more English-language tabloid dailies for niche markets: *Project Eyeball,* which was launched in August that year, was aimed at the Internet generation; *Streats,* which debuted a month later, focused on commuters. *Streats* published about 230,000 copies a day and *Eyeball* about 200,000. (In June 2001 SPH announced the closure of *Project Eyeball* because of a slowing economy.)

After the laws changed to permit ownership of print and broadcast operations the Media Corporation of Singapore (MCS) launched *Today,* a tabloid daily. Until then MCS was solely a broadcaster: Two of its business units operated five television and 10 radio channels, comprising the bulk of Singapore's broadcasting. MCS put up 49.9 percent of the $7 million to fund *Today,* RFP Investments (part of mass rapid transit operators SMRT Corporation) 30.2 percent and Yellow Pages 19.9 percent. SPH decided to take MCS head-on and in mid-2001 announced plans to launch two television channels. The English-language channel was called TV Works, and its Chinese-language counterpart was called Channel U. SPH hired consultants from Reuters and the BBC to help prepare for multiple-platform journalism. SPH trained many of its print journalists to supply the content for the television news and current affairs programs (personal observation 2001). Journalists moved into a purpose-built site to house both television and newspaper production late in 2001. A high-stakes battle for audience subsequently took place between Singapore's two major media groups, as each ventured in different directions down the multiple-platform path. In terms of convergence, we should note the impact that changes in the law had on Singapore's media structures and relate those to other environments.

In June 2003, the US Federal Communications Commission (FCC) approved controversial changes to the rules governing media ownership, which potentially influence the spread of convergence in the US. Limits on ownership of newspaper and television have not been updated since the mid-1970s, when cable television was still in its infancy and the Internet was even younger. FCC Chairman Michael Powell and his two Republican allies won the vote against the protests of the commission's two Democrats. The proposed new rules, yet to be implemented, allow media companies to own more outlets in the same markets. Several groups appealed the decision in various federal courts. These cases were consolidated and assigned by lottery to the US Court of Appeals for the Third Circuit.

In terms of convergence, the proposed changes would raise from 35 percent to 45 percent the cap on a single company's reach within the national broadcast TV audience. Rather than limiting the number of television stations a broadcaster can own, the FCC has limited the number of households a broadcaster can reach. The changes also ease the limits on owning more than one TV station in a market; ease restrictions on owning both a newspaper and a TV station in the same market in the majority of large markets; and they ease restrictions on cross ownership of radio and TV stations in the same market. Early in 2004 Congress rolled back the single-reach cap from 45 percent to 39 percent, removing this part of the Third Circuit's review of the proposed changes. At that time, 39 percent was the maximum reach by any one network. The biggest audiences were: Viacom (CBS and UPN) with 39 percent of households; Rupert Murdoch's Fox network, 38 percent; General Electric's NBC, 34 percent; and Paxson, 32 percent.

The FCC, established in 1934, is an independent government agency responsible for regulating US radio, television, satellite, and cable broadcasting. Traditionally, the FCC has sought to prevent the emergence of media monopolies by establishing limits on ownership. The June 2003 decision was the culmination of a review process that began in September 2001 when FCC chairman Michael Powell created the Media Ownership Working Group and charged it with developing a foundation for re-evaluating media ownership policies. The FCC said it hoped to develop policies that promoted competition, diversity and localism in modern media markets. It was the most extensive review since the 1996 Telecommunications Act required the FCC to examine media ownership rules biennially (FCC 2004). Any new rules will not be implemented until the Third Circuit makes a decision, which had not happened prior to the writing of this book. In a long analysis, the London-based *Financial Times* reported that if the changes proceeded despite the opposition in Congress, smaller US networks could merge and major newspaper groups such as The New York Times Company and the Tribune Group were likely to buy television stations in markets they were currently excluded from (*Financial Times* 2003).

Gil Thelen, publisher of the *Tampa Tribune* in Florida, noted that the regulatory confusion—what he called "the judicial quagmire"—had clouded the situation regarding the spread of convergence as of early 2004. A lot of change had been held up by the appeals, he said, and it would probably take a long time to be resolved (2004). Andrew Nachison, director of the Media Center at the American Press Institute, said the model that Media General in Tampa had devised was the most likely to be replicated in the US if the FCC rule changes permitted one company to own multiple media in one market. But he maintained that convergence would happen with or without the FCC changes. "Print, broadcast and online media will find ways to partner, share resources and extend their reach because audiences are dispersed across multiple media. The strategy is motivated by [changes in] the audience, rather than the device," he said (quoted in Glaser 2004b).

Elsewhere in the world, legislation forbids or hampers editorial managers' aspirations toward convergence. In the UK, the strict regulations governing cross-media ownership have significantly hampered the plans of managers hoping to integrate editorial teams (Pascual 2003a: 34). In Australia, the issue of changes to media ownership laws has been around since the end of the twentieth century. In essence, the Liberal-National Party Coalition that has been in power since 1996 supports law changes that would allow one person or company to own a newspaper and a television station in the same city. The Labor opposition wants to maintain the status quo. The Hawke-Keating Labor government had introduced a law in 1987 that forbade any one proprietor to own more than 15 percent of both a newspaper and television station in any one metropolitan market. The then treasurer Paul Keating famously said, "You can either be a prince of print or a queen of the screen." In October 2000 the then communications minister Senator Richard Alston declared that the coalition government had gone through too much "pain and grief" when it tried to change cross-media ownership laws without the support of the Labor opposition. In July 2004, in the lead-up to a federal election expected that year, opposition leader Mark Latham said that Labor would ensure the media ownership laws did not change if it gained power. "We support the cross-media laws that maintain different owners for TV stations and newspapers in a capital city and I think that is the best thing we can do for diversity," he said. "These are the laws the Howard government wants to get rid of, but it is Labor policy to keep them" (quoted in O'Loughlin 2004: 5). Australia presents a useful case study of the influence of laws on media structures, and the situation is discussed in more detail in the next chapter.

Macro to Micro: Factors That Make Convergence Possible

This section of the chapter moves from the big picture of social, technological and legal change to the smaller world of implementing media convergence. It looks at the factors that inhibit and foster convergence within individual organizations. Until mid-2004 James Gentry was dean of the William Allen White School of Journalism and Mass Communications at the University of Kansas. He relinquished the position to concentrate on teaching and research, and to consult with media groups seeking to implement convergence. Gentry proposed a continuum between "easy" and "difficult" introduction of convergence, noting that the quotation marks around the words were needed because "there really is no such thing as easy convergence" (Gentry 2004). Figure 2.1 lists the two extremes of the continuum, and these factors are described later in the chapter. Note that at times the continuum can be quite wide.

Declining economic conditions and the presence of competition in the market contribute to the potential for change. Mitch Locin, senior news editor for electronic news at *The Chicago Tribune*, said the *Tribune* felt competition from everywhere: "TV, radio, cable, local papers both dailies and weeklies, and big papers like *The Wall Street Journal* and *The New York Times* that get home delivered, plus online delivery of major

Figure 2.1: "Easy" versus "difficult" convergence

"Easy" convergence	"Difficult" convergence
Central to organization's strategy	Not central; secondary or worse
Committed and focused leadership	Other leadership priorities
Culture of innovation and risk taking	"Always done it this way"
Coordinating structure	No coordinating structure
Same ownership	Different ownership
Same values	Different values
Aligned systems and processes	Systems not aligned
Partnerships with Cable television	Partnerships with over-the-air broadcaster
Past successes together	Previous problems or no relationship
Cultures flexible or similar	Cultures not flexible or similar
Co-located	Located some distance apart
Absence of labor unions	Presence of strong unions

news organizations like CNN.com, MSNBC.com and NYTimes.com. Even in a market where you may be the only newspaper in town, that's not really the case any more with the web" (Locin 2003). Factors unique to each specific market and culture, such as the dominant demographics of a region, can be significant. John Burr, assistant managing editor at the *Florida Times-Union* in Jacksonville, said the number of retirees in the state made a difference to the adoption of convergence. "You have a lot of people from around the country retiring here, and older people read newspapers," Burr said. "The people coming to Florida have money. That's why newspaper chains have coveted this state and put money and effort into doing it well. That goes back 40 years. Why are we so convergent? The newspaper chains have built up their best newspapers in Florida, they're full of pretty creative people who really want to do this and they're pretty competitive" (quoted in Glaser 2004a).

One aspect of having a creative staff was the ability to remain competitive and to have a flexible mindset. Locin of *The Chicago Tribune* noted that the paper used to publish an afternoon edition in the mid-1990s. "We're publishing a midday edition now, just like we used to, only now we're putting it out on the web," he said. "So reporters have to file every day by 11[A.M.]." Newsplex training director Martha Stone said the process of moving from a mono-media to a multi-media company involved change management, but that had not been part of the process at some companies. The result had produced "a lot of shocked, unhappy people." Stone called for more communication and "more sensitivity to how practices change and how humans respond to that change." Why did some people resist change, Stone asked. "Probably because they have not been communicated with and because there is not enough care injected

Why and How Convergence Is Emerging

into the transition process" (quoted in Roper 2003a: 30–32). Appropriate change management processes and intelligent selection of staff produces people with the flexible mindset that convergence requires. Chapter 8 discusses these factors. Media groups also need to be willing to train their staff, to prepare them for new roles and new environments.

In May 2002, Ifra's then director of editorial strategy Ruth de Aquino presented a report on the evolution of convergence in Europe. Funding for the research, conducted by the MUDIA (multi-media in a digital information age) consortium, came from the European Commission. The report concluded with a list of essential requirements for companies seeking to instigate convergence. That list is reproduced here so readers can match it against Gentry's continuum:

* A top-down management approach that focused on a multi-media integration philosophy.
* Training in multi-media that allowed news professionals to develop a new and diversified style of gathering news.
* Marketing development of a multi-media brand.
* A mix of multi-skilled journalists and specialists so that they could help each other.
* A single staff creating news content 24 hours a day for all media platforms as an integrated process, from production to delivery.
* An editorial system that integrated all platforms of the company with a common database for all to access.
* A physical space, shared by journalists from all media, which should contain a multi-media desk, and ideally a radio studio and a television studio.
* Integration between technical and editorial staffs for constant updating of the editorial system according to the needs of the newsroom.
* Focus on how to complement each other during news meetings throughout the day.
* Multi-media packages for advertisers.
* Cross-promotion amongst all platforms.
* Opposition of laws that hinder the development of the media industry.
* Identification of strategic partners to make it easier to diversify (de Aquino 2002: 13–14).

The next section of this chapter discusses each of Gentry's criteria and relates them to convergence developments around the world. It also incorporates the ideas in de Aquino's list above.

Organizational Strategy

For convergence to work, it must be central to the organization's strategy, and not merely a reaction to the latest industry trend. This means allocating an appropriate budget and appropriate people to run it. In 1991 Dolph C. Simons Jr., who had been editor or publisher of the *Lawrence - Journal-World* since 1962, gave a major speech to mark the 100th anniversary of the Simons family's arrival in the news business in Lawrence, Kansas. Dolph C. Simons Jr. is the grandson of W. C. Simons, the company's founder. The World Company is one of the most converged media groups in the world. "We believe it is important to look upon our business as an information business," Simons said, "[and] not merely a newspaper or a cable television operation. We want to stay abreast of new developments and be able to deliver news and advertising, as well as other information, however a reader or advertiser might desire." Many researchers cite Arthur Sulzberger Jr.'s similar quote ("if people want information beamed directly into their minds, we will create a cerebral cortex edition") as evidence of the New York Times Company's commitment to convergence. Both are indications of the clear organizational strategy allied with focused leadership so key for convergence to work, though the Lawrence quote was well ahead of the New York equivalent.

Rob Curley, director of new media and convergence for the World Company, said the "number one thing" for convergence to succeed was the need for "ultimate buy-in from the highest level" of the company. "Otherwise it does not work," he said. Commitment from the top made his role easier in the sense that he had clear directions, and he could then expect commitment from his staff. "You are going to change the people, or you are going to have to *change* the people. They are going to have to buy in or you are going to have to make changes. You cannot have a cancer in the newsroom." Curley said his company rewarded people who got involved with convergence (Curley 2004a).

Editorial managers at the *Orlando Sentinel* thought strategically about the process of introducing convergence. They looked at how people received news and how audiences were changing and fragmenting, and reflected on how the *Sentinel*'s share of advertising dollars in the market would change and what that could mean long term. The then editor, John Haile, said managers emphasized from the outset that it was vitally important to reach more people. "New competitors are coming at us every day. We have to take our journalism to where people are. We've got resources that no local competitor can match, but we have to use them or risk losing them." Money had to be found to sustain good journalism:

"We made it clear that this wasn't a lark—that we had to try to figure out how to change the business, because if we didn't we weren't going to have the resources to finance this operation. If we don't keep finding new audiences, and start losing revenues, then this news organization is going to get crunched." The changes reflected Haile's thinking about the future of newspapers. "We spend too much time worrying about how this new world will affect how we do journalism," he said. "We've got to have confidence we can practice good journalism across multiple media. We can't get tied in knots about things" (quoted in Gentry 1999: 7).

Committed and Focused Leadership

Classic innovation theory shows that leaders are vital factors in delivering change. Gloria Brown Anderson, vice president for international and editorial development at *The New York Times*, described how in 1992 managers produced a statement that described the paper's philosophy in 13 words: "Editorial excellence and independence are essential to our profitability and profit sustains them." As the process of cooperation extended through the company, managers hammered out—in another 13 words—the company's core purpose: "To enhance society by creating, collecting and distributing high-quality news, information and entertainment." Chairman and publisher Arthur Sulzberger Jr. said the *Times* had to be "agnostic about the means of delivery, be it ink-on-paper, television, radio, beepers, cell phones, portable data assistants or other media yet to be invented." A commitment to editorial excellence and independence put the paper's editors at the "heart" of the newspaper and those editors were challenged to create journalism that would accomplish the goal of "best enhancing society." All stages of the communication process had to "emphasize the primacy of the editorial content" (Anderson 2001).

The Chicago Tribune's Mitch Locin also emphasized the role of communication. One of his main jobs was to work with reporters, photographers, and their editors to explain what was needed. "It takes time. There's no getting around it, it takes time out of your day. I explain things." Locin was a former political reporter who had covered presidential campaigns, so he had credibility in the newsroom. He said he knew how much time some reporters wasted "kicking things around" and leaving the building for a cigarette or coffee. "So 10 minutes for [doing] a TV interview is really not cutting into a reporter's day that much," Locin said. The paper's reporters needed to understand that they were "part of this revolution"—that finding content for all platforms was a key part of the job. Managers were responsible for making reporters feel that they were a part of the convergence process. "In the best newsrooms that comes from the top or at least from reporters' bosses" (Locin 2003). Thelen also emphasized the importance of communication: "It's vital. Unless you have the interpersonal organizational skills in high working order supported by the appropriate technology this stuff does not get very far. Because our best work has been where—in an open, trusting way—we address problems, brainstorm, try things out on one another and come up with a better answer than each of the platforms or their leadership could have done alone" (Thelen 2004b). *Tampa Tribune* multimedia editor Ken Knight said it was essential to get buy-in from the very top. "Because staff look for guidance, it is important that managers and editors have a clear picture of where they want to go with convergence" (Knight 2004).

The Dallas-based Belo Corporation owns *The Dallas Morning News*, WFAA-TV Channel 8, cable channel TXCN, and DallasNews.com. Newsroom executives from each newsroom meet regularly to talk about projects and attend meetings at the newspaper, television, and online sites. Teams consisting of people from all media work on stories and projects together. Each newsroom shares its daily list of stories to be covered with other newsrooms via a database that includes complete stories from WFAA, TXCN, *The Dallas Morning News* and all websites. Each also emails stories and scripts to journalists and managers at Belo's Dallas properties several times a day. David Duitch, vice president for news at WFAA, said "sharing makes all sides stronger and allows us to respond faster to breaking stories" (quoted in Stone 2004: 38). Newspaper consultant Martha Stone said the successful transition at the Dallas-based Belo Corporation from a culture of competition to cooperation was the result of "a well engineered and well communicated top-down management commitment to cross-media integration." John Granatino, vice president of interactive sites for Belo Interactive, credited the strategy shift to publisher Jim Moroney. "Without Jim here, there would be none of this," Granatino said (quoted in Stone 2004: 37).

The chief editor of the Ming Pao Newspaper group in Hong Kong, Paul Cheung, similarly said the move to a multi-media environment could only be successful if accompanied by a corresponding change in the attitude of editorial managers: "The chief editor . . . must be a leader in terms of the changes." Leaders were needed to alter attitudes from the "old" mindset to a multiple-media approach. Cheung and his staff had several brainstorming sessions and visited newsrooms at *The New York Times*, *USA Today*, the BBC, and *The Financial Times*. Senior staff spent one- and two-week attachments in those newsrooms. "We decided the integrated approach was more feasible in Hong Kong. It would be both cost effective in terms of saving money and it would prepare us for the future. The integrated newsroom is the newsroom of the future" (Cheung 2001; quoted in Loh 2000: 48). Researcher Janet Kolodzy quoted an anonymous journalist who said that reporters needed "managers who know the value of all media so that this new tactic can be harnessed properly" (Kolodzy 2003a: 61). *Orlando Sentinel* editor John Haile understood the importance of committed and focused leadership. "We repeatedly made the argument [to staff] that the greatest risk to a strong journalistic enterprise was that revenue streams would begin to dry up, meaning fewer resources for the newsroom." In other words, he kept people focused on the reasons for change. Because of this policy, Haile noted his staff had "intense interest in finding ways to move with the audience" as technology changed the way people received news and information (2003: 10).

Culture of Innovation and Risk Taking

The Poynter Institute's Howard Finberg maintained that the big issue within media organizations was the need to focus on learning and adjusting the characteristics of the entire organization. "With education we can affect the learned behaviors of the media industry's leaders, its journalists, and other workers. And when I talk about the media industry leaders, I am not talking about just the people who sit in the boardrooms. Leadership includes managers and staff members, who actually can be more influential than their bosses." Finberg said it was necessary to change the behavior traits of members of the media "society" who avoided risk and constantly opted for the short view. "We need to change this cultural template so that the media industry adopts a vision that includes the long view." Part of this was investing in training and education; another part was fostering innovation (Finberg 2003).

An understanding of the role of "culture" is a key to the transition to convergence. Much has been made of the differences between print and broadcast journalists. Broadcasters regard print reporters as lazy because the latter spend their day with contacts and only produce one story a day, often waiting until late in the afternoon before they start writing. Print people dismiss broadcast folk as shallow, seeing them as poor spellers who skate over the surface of events and write "headline news." Former *Orlando Sentinel* editor John Haile said that in the meetings held to discuss a potential relationship with a television station, the staff expressed "concerns about all television news" and questioned whether the newspaper should get involved. "They saw local TV news as superficial, often fluffy and not particularly well reported. They said that, generally, they had little respect for the work of television journalists and even less for the news anchors. The newspaper's journalists simply believed their standards were higher and their focus much stronger on news of importance to the community" (2003: 7–8). An anecdote from *Orlando Sentinel* reporter Scott Maxwell distills this notion. He described the first day of his internship at an unnamed television station: "We unloaded the equipment and the television reporter went over to a newspaper rack, took out a paper and said: 'This, my friend, is your research.' The reporter was counting on the newspaper to do all the research. I decided then that this was not part of the business that I wanted to be part of" (Maxwell 2004).

But cultural differences are learned. Howard Finberg pointed out that much of human life was transmitted genetically—a baby's desire for food, he said, was triggered by physiological characteristics determined within the genetic code. But people learned culture. "My desire for eggs and bacon, however, is not a genetic characteristic. Rather it is a learned (or cultural) response to morning hunger." If culture was learned, it could be re-learned, he said. Researcher Jane Singer said much of the misunderstanding between print and broadcast journalists was cultural and could be managed through "appropriate management actions." She studied newsrooms where managers had required staff to work together, promoting "synergy" by "emphasizing the need for reporters and editors to share information with previous competitors across media platforms, hoping the trust that comes from such sharing will grow one pair of journalists at a time." She found that newspaper journalists enjoyed working with counterparts in other media, and the fact that each had gained respect for the other group "lends credence to this strategy" (Singer 2003a).

The *Tampa Tribune*'s Thelen had discovered this key: "Cultural resistance is the biggest hurdle for converging newsrooms," he wrote in the industry journal, *Quill*. "For multi-media work to take deep root, journalists from once-competing newsrooms must learn to cooperate and collaborate—a tall order in our highly individualistic professional mystique" (Thelen 2002b: 16). The Poynter's Finberg, a former newspaper executive, came to the same conclusion, though from a different direction. "Understanding our culture is the first step in trying to change it. Learning new behaviors and sharing those new experiences across the media industry's society will be fundamentally important. Getting the industry and the boardroom to embrace convergence is about getting the industry to understand what's at stake and acknowledging our own history of failed attempts at suppressing new communication technology. By learning from our past, we can prepare ourselves for the future" (Finberg

2003). Newspaper consultant Dr. Juan Antonio Giner said that the recent history of newspapers suggested an "adversarial tradition" that resisted integration. Some of the conflict was between "journalists and managers, editors and reporters, writers and layout people, technicians and newsrooms, marketing and journalism, PCs and Macs, news and opinion, information and entertainment, and newspapers and television." Digital technology and the integrated journalism it made possible were powerful agents for elimination of many of these conflicts (Giner 2001b).

A Coordinating Structure

Human beings and the great apes share almost identical DNA, but they are quite different creatures with different abilities. This is because they have radically different skeletons, and these differences in structure influence their ability to perform tasks. Structures similarly dictate how well or badly convergence can happen. Structures can be put in place to influence the acceptance of convergence in the newsroom. At the *Ming Pao* in Hong Kong, chief editor Paul Cheung aimed to involve all his staff. Between June and December 1999 in preparation for the move to convergence journalism, Cheung held weekly meetings with senior editorial staff. Cheung used the meetings to sell new ideas. Because of the organization's culture, these people passed the message on to their staff. Communication reduced uncertainty. The *Tampa Tribune*'s Thelen and Cheung also said they built convergence involvement into individual reporters' annual performance appraisals.

Figure 3. The newsroom of the *Orlando Sentinel* with the TV newsroom in the distance. Photo: Stephen Quinn.

Locin of *The Chicago Tribune* also noted that a reporter's willingness was discussed during the performance review. "It's up to individual bosses, mid-level editors, to hold their staff to account." Locin said he kept "plugging way" because convergence was "an evolution rather than a revolution," noting that it would "take time for it to become second nature." Locin drew parallels with the introduction of computers into the *Tribune* newsroom to replace manual typewriters. "Some reporters were typing on them as they [the typewriters] were being carried out the door. It takes a while for people to adjust" (2003). Part of the structural issue involves telling reporters that they are required to take part. As convergence became more ingrained in the *Orlando Sentinel* newsroom, so did the expectation that everyone would participate. In the late 1990s reporters had the choice of whether to work in television. The managing editor at the time, Jane Healy, said within a few years anyone interviewed for a job was expected to appear on television. "The standards have slowly evolved to where we just expect it of everyone now" (quoted in Gentry 1999: 5).

Common Ownership

Ann Gardner, the former managing editor for multi-media at the World Company in Lawrence, said having one owner was a significant factor in the transition to convergence. "We were in a really good position to do this because we already had the TV station and the online operation with the newspaper [under one owner]." The company moved into a custom-built newsroom in September 2001, partly because of limited space in the old newsroom. Having the same owner made the process relatively easy, both in terms of the move itself and in reaching the decision to move. The single owner was also a factor in terms of negotiating for the future: "[W]e said if we're going to build this new space we need to look down the road and it was obvious that there would be more convergence media playing a role, so it was a natural move" (quoted in Roper 2002a: 38). Gardner's successor, Rob Curley, said the success of convergence at the company—as of early 2004 the newspaper had recorded 18 straight months of circulation gain—was a direct reflection of the fact the company was owned by one family. "It is very progressive family ownership, so progressive that they began laying cable in 1968. It was one of the first cable companies west of the Mississippi." The company received a lot of criticism for this move, Curley said. "Now I think Lawrence has the number one cable penetration of any city in the country. It also has the number one high-speed cable modem penetration" (Curley 2004a). Curley said the Simons family disliked complacency and were always telling people "we must do better tomorrow compared with today." The idea of convergence was not some snappy buzzword for them, he said, "but a way to serve the community" (Curley 2004a).

Forrest Carr, news director at WFLA in Tampa, admitted that successful convergence was difficult. It was not necessary to have the same owner, he said, but it certainly helped remove a number of serious roadblocks and made it possible to drive convergence from the top down. "Having each platform owned by the same company also removes one of the biggest obstacles among potential convergence partners—the idea that each partner must always benefit equally from convergence. Holding on to that notion is one of the surest ways to derail convergence. It's the viewers, users, and readers who should benefit the most." The same ownership helped maintain the cooperation needed for achieving the goals of convergence, Carr said. One of the most important of those was the notion of "better serving the public through stronger journalism" (Carr 2002b).

Most big media groups such as the Tribune Company and Media General have separate broadcast, print and online or interactive divisions. (The Tribune Company owns 11 newspapers, 26 television stations and operates more than 50 websites. Media General, which owns the *Tampa Tribune,* tbo.com, and WFLA-TV in Tampa, has a total of 26 daily newspapers, almost 100 weekly publications, a total of 26 television channels, and more than 50 websites.) It could be argued that because of corporate issues such as political in-fighting and arguments over financial control, separate divisions would work against convergence. But in general the larger a corporation, the greater the need for separate business units to manage the company. Media General's Tampa network is a manageable size. The *Tampa Tribune*'s Thelen said he set out to align the three platforms to think about the market and market share, both in terms of advertising and audience, with the aim of growing revenues and audience. The three media platforms in Tampa also have someone to provide an overview, a form of helicopter vision for the company. In August 2003 Ronald Redfern was named president of Media General's Florida Communications Group. Redfern focused on the proper integration of the three platforms. "This has been a vital step in the process of learning to think as one," Thelen said (2004b), because multiple-media companies needed to have "a single person thinking about the totality of the effort" (Thelen 2004a). Smaller companies like the World Company in Lawrence are better positioned to be innovative because they are nimble and easier to steer. Regardless of the overall company structure, having the same owner or manager for all platforms makes management easier. It decreases the possibility of variations in values which can lead to staff pulling in different directions. At the editorial level, innovative companies have appointed individuals to run specific departments for all platforms. In Tampa, Rick "Duke" Maas is the senior editor for all sport departments.

Shared Values

Curley of the World Company said other companies were doing convergence because the return on investment was positive, or because it meant profits for stockholders, but his company had adopted convergence because it was good for the town of Lawrence, Kansas. The approach of the Simons family was to ask, "How can we serve the community?" he said. "It's subtle because the bottom line is still the same thing. . . . We're doing convergence not as a way to get more work out of people, but because we think it's the right thing." The company had aligned its values to those of the community (Curley 2004a). Mindy McAdams, a professor at the University of Florida at Gainesville, said that for convergence to happen staff had to think in terms of teamwork: "[T]hey should say our news operation, our TV news, our paper, our website. They should say 'our' not because the corporation tells them to but because they believe it—or better, they know it. They should try to get every story out on all their platforms, as appropriate, and always do a good job of telling the story" (quoted in Glaser 2004b). The man who initiated convergence at the *Orlando Sentinel,* John Haile, said emphasis on shared values was at the heart of the changes. "We had to emphasize basic journalistic values. We made a point to emphasize that there were fundamental values that will define us, no matter what medium we are in. These values set us apart from the competition. They are our competitive advantage." Prior to the changes, Haile organized a series of discussions about fundamental news values. "When we got involved with other media, we didn't want to be dumbing down what we do. We wanted good journalism to carry across other channels" (quoted in Gentry 1999: 4).

Haile has continued to emphasize the importance of sharing values and information: "To help spread the message . . . , the *Sentinel*'s new publisher John Puerner, the advertising director and I conducted a series of meetings designed to reach every person at the company with the message that the *Sentinel* was intent on becoming a multi-media information company" (2003: 10). With the passage of time, convergence flows through an organization. In the right environment of similar values, trust, and communication, it can generate a momentum of its own. Ken Knight, multi-media coordinator at the News Center in Tampa, said that after four years in the new location, convergence was happening on many different levels. "Two reporters or two photographers could be talking about how best to do a story. Or two editors could be sharing information." He said it would be a mistake to try to make convergence funnel through one person because that ignored the vital aspect of sharing between reporters. "As long as we are creating an environment in which the information is important to the community, then we all win" (Knight 2004).

Aligned Processes

Tampa's Gil Thelen likened a converged news operation to "a full orchestra" that over time demonstrated to its citizens and customers that the content they received was authentic and reflected the community. A symphony has one conductor. Does a media grouping such as the News Center in Tampa, with three distinct media platforms, need a single conductor? Thelen argued that Ron Redfern performed this role. "His focus is less on the newsrooms than it is on the proper integration of the three organizations." Thelen said convergence needed "a single person thinking about the totality" of the process. "Whether that will take us to an editor-in-chief on all platforms, I'm not sure. That would have to be a very special person. You have to work within the human limitations" (Thelen 2004b). Convergence also needed technology that made sharing of information, contacts, story ideas, and content easier. The News Center was experimenting with a range of technologies to aid this process. One of those was NewsGate, which helped integrate budgets on different platforms.

A multi-media desk represents a physical symbol of convergence, and should be promoted as the hub of convergence, but it is also necessary to generate psychological manifestations of integration. One of the most simple but powerful of these involves publicizing successes. Newspaper managers should reward reporters who produce excellent content. In Australia, *The Age* and *The Sydney Morning Herald* newspapers circulate in-house newsletters in full color that feature the achievements of multi-media and traditional journalists. Sometimes minor awards are appreciated. The editor of orlandosentinel.com, Anthony Moor, said to generate rivalry in the newsroom his company created a "Golden Mr. Potato Head" award for the best online story of the week (ONA 2003).

Aligned Technological Systems

Ruth de Aquino said the absence of appropriate technological systems to link newsroom budgets was one of the key obstacles to convergence in Europe. Newspapers needed a common database to hold information for all media (2002: 13). Curley said his staff had to customize software to coordinate news stories at the News Center in Lawrence. "We had to write our own budgeting software because no budgeting software out there would allow all three media to share budgets. So we built it ourselves. We don't have a name for it. It's not sophisticated but it allows all three mediums to see what all others are doing" (Curley 2004a).

Similarly, the News Center in Tampa built a common software system called BudgetBank, soon after moving into a dedicated building in 2000, to ensure all three platforms knew what stories were being covered. Multi-media coordinator Ken Knight said an earlier multi-media editor, Steve DeGregorio, created the first version. Allyn DeVito, one of the news technology managers, devised the version in use as of mid-2004. It is the only system that all three platforms in the News Center have access to. BudgetBank is intranet based, and icons indicate the type of story being planned. The icon "c" means a convergence story; a "TV monitor" icon represents a television package; a "camera" icon meant a still photographer would also shoot video; and the "NB" icon next to a story indicated material that would be shared with other Media General publications, such as a news agency story. Knight said the News Center was working with CCI Europe, a Danish software company, to develop a system for coordinating print and broadcast systems (Knight 2004). NewsGate was being trialed as of late 2004. Publisher Thelen pointed out that the management of information was dimensionally more complicated in a converged environment. "The demands on editors and producers increase exponentially" (Thelen 2004a).

Common Databases

Twenty-first century journalists work with information as their core material, so they need quick and reliable access to it. Databases provide one efficient way to assemble and manage information. Databases offer several advantages. Information in a database can be extracted to use in different media, and digital tools allow information to be transferred relatively easily from one format to another. Databases also remove the need for paper and photographic archives that take up valuable space. The Spanish newspaper *La Vanguardia* has introduced several databases to promote future technological development. José Luis Rodríguez, the paper's deputy editor-in-chief, said digitalizing, indexing, and filing the newspaper group's contents and information saved time when reporters needed to find archived material. It also freed up space in the newsroom. The photographic process was completely digitized, he said, which had made access to content faster and more efficient. It also offered the possibility of selling images to third parties and publishing them immediately on the group's websites.

In Sweden, the national newspaper *Aftonbladet* used databases to manage the distribution of audio and video content between its print media and websites. Bella Tidblad, the company's editorial development manager, said it was important to gain journalists' trust and increase their multi-media knowledge. It was also essential to implement technologies that allowed the group to create content for different media and to transfer it easily from one platform to another. "Our challenge now is to allow content to flow in both directions—from print to web and back—and to ensure each news item is adapted and published in the best possible way for a given outlet" (quoted in Pascual 2003b: 49).

Cable Television Partnerships

Network television channels by their structure are part of media conglomerates, whereas cable television channels tend to be smaller and nimbler. This gives them the ability to make relatively quick decisions, and they tend not to be obligated to shareholders to the extent that network channels are. Gentry said convergence was more likely to emerge because of this flexibility. Ann Gardner, former multi-media managing editor at the World Company in Lawrence, Kansas said convergence had had more chance of success when initiated because of past links: "Because of the ownership situation we've been crossing over a little bit for probably 15 years." It also helped to have "a smaller ship" to move compared with some of the larger publishers. The single ownership "helped a lot," Gardner said. "We have only one set of owners that we have to respond to and that is nice. Although we encourage people to keep a competitive mentality between the staffs, it's not like having people who are literally financial competitors. If you had two entities that were really competing with different owners for advertising dollars, I would think that would be a much more difficult negotiation" (quoted in Roper 2002a: 39).

Appreciation of Different Cultures

Joe Brown, editor of the Pink section of the *San Francisco Chronicle*, was intimately involved in the convergence negotiations between KRON-TV and his paper. "We needed to educate everyone within the newspaper how important it was to have relationships with television and online. We had to teach everyone within the newspaper to understand why convergence was important." Brown appreciated the importance of recognizing culture differences, and organized a visit to the television station early in 2003. "We went over there. We took a road trip to visit them. We needed to develop trust and understanding. So we packed our people up and sent them over to the TV station. We had lunch with them. It broke down a lot of the barriers" (Brown 2004). The paper's associate managing editor, Kenn Altine, said it made a huge difference when top management understood both television and newspapers. "They do not have to be involved with the daily management, but they have to understand the differences between the various media. You have to know what you want and you have to cooperate. They [newspaper managers] have to understand that there is a skill involved in working in television." Editor Phil Bronstein appeared regularly on television and radio, and understood the requirements of each media (Altine 2004).

Writer and commentator Mark Glaser is well aware of the cultural suspicions: "TV people might think newspaper people aren't fit for going on-air, and their print reports obviously need to be doctored for broadcast. Newspaper people might think that TV people are simply blow-dried actors reading lines. And the web people are viewed by everyone else as just nerds in the back room" (Glaser 2004a). "The toughest hurdle is culture," Anthony Moor of orlandosentinel.com said. "True convergence, where content sharing is standard practice, requires a subtle understanding of the needs and methods of each medium. So I spend a lot of time seeking to bridge the gap through education and outreach. I say I'm a web editor, but I'm really a change manager" (quoted in Glaser 2004a).

Tampa Tribune publisher Gil Thelen described the preparations for convergence at his organization in terms similar to the evolving relationship between a courting couple. "Probably the most important thing we did before we moved in together was to get to know each other. We dealt with the whole values question. What is public service? What are your principles of news judgment? What are the roots of our craft? We figured out we shared fundamental values: things like public service and balance. That has made the News Center work." If WFLA had produced tabloid-style news, Thelen said, convergence would not have worked (Thelen

2004b). WFLA news director Forrest Carr said he had found much more common ground than he had thought possible at the News Center. "Our journalism styles were not too far apart to begin with. The partnership would have been much more difficult, though still not impossible, if WFLA-TV's news style were tabloid and sensational in nature" (Carr 2002b).

Common Location

Because of the importance of communication, convergence works best in situations where people are close enough to each other to be able to talk face-to-face and meet regularly. Proximity reduces confusion. "Geography is destiny," Tampa's Thelen famously noted in 2000. Proximity means people can share information more easily; with time, proximity also generates ideas and trust. "One of the things that made convergence possible in Tampa was the fact that we live together," Thelen said. "We eat lunch together, go up in lifts together. If you do not have this proximity, you have to do it with technology" (Thelen 2004b). Multi-media coordinator Ken Knight said: "I talk to people all the time. It's wonderful to be able to have [numerous] face-to-face conversations because it gives a stronger sense of understanding." The three media platforms were converging informally when they were in separate buildings, Knight said, but integration then depended on editorial leadership. "It did not really percolate into the reporters' fiber as much as it does now." When all platforms moved into the Tampa News Center in 2000 it created a synergy. "Being in the same building gives us a sense of urgency and immediacy" (Knight 2004).

Curley described the News Center in Lawrence as his company's "ode to convergence." It reflected the value the company placed on convergence and its staff. The building is indeed magnificent, and all staff this author spoke to said it was a pleasure to work in (personal observation 2004). (See chapter 5 for more details.) Curley said editors believed it was important for journalists to talk to each other. "The newsroom is not broken down by media, but by beat. The cop reporters for TV and print sit next to each other. So do the education reporters and the council reporters. The sports folks for the newspaper, 6News and KUsports.com all sit next to each other." The newsroom also produces five to seven television programs, on subjects such as homes, gardens, and local music. "Some of the people who produce the local origination programs sit with the [newspaper] features staff."

Such was the nature of the work environment at the News Center, Curley said, that the newsroom rarely had vacancies. "When people come here they stay. We currently have our first opening in 18 months, and the applications we are getting are mind-boggling, from the top journalism programs in the country. People come here for the chance to do world-class journalism. World-class journalism has absolutely nothing to do with how many newspapers you own" (Curley 2004a). The chief editor of Ming Pao Newspapers in Hong Kong, Paul Cheung, led the move to convergence at his Chinese-language daily by integrating many of the teams on the newspaper and online editions. Business reporters at the online and newspaper sections, for example, have a combined editor and sit together (Cheung 2001).

Absence of Labor Unions

In her May 2002 report on the evolution of convergence in Europe, de Aquino said the two main "external enemies" to convergence—mentioned mainly by French and English newspapers—were unions ("rigid labor laws, contracts and function definitions") and legislation that restricted cross-media ownership (2002: 7). The website of the American Press Institute publishes a "convergence tracker" that details the number of converged media operations in each state. As of mid-2004 Florida had eight converged groupings, almost double the number of any other state (the next nearest was Texas, with five). Anthony Moor, the editor at orlandosentinel.com, said the absence of unions contributed to the development of convergence. "When I was on the union side in San Francisco [at KRON-TV] our union clout slowed convergence," Moor said. "In Rochester [at the *Democrat*], I was on the management side, and we learned about reasonable union concerns but also saw unreasonable fear; that, too, slowed things down. In Orlando, we don't have unions, but we still have to be careful about what we ask our staff to do, because if they don't buy in to change, it won't happen" (quoted in Glaser 2004a). Asked why so much convergence was happening in Florida, the assistant managing editor for multi-media at *The Los Angeles Times*, Joseph Russin, said "no unions" was one reason, adding that newspapers in Florida did not seem to be so "hide-bound" (comment at ONA 2003).

A consultant for the World Association of Newspapers once defined convergence as "the art of getting journalists to do more." It reflects a major concern of editorial managers and publishers: How to pay for the major developments needed to do convergence. The next chapter looks at the revenue and business models for multiple-platform publishing. It is important to consider these financial aspects because, as Ifra's Kerry Northrup has often pointed out, convergence is more a strategy for growth rather than a way to save money.

3 Business and Revenue Models for Convergence

The business processes and potential income models associated with convergence are the subject of this chapter. It also considers convergence as part of the change management process, and looks at the revenue models available to modern media organizations. The chapter ends with discussion about the trend toward multi-media advertising. In particular, this chapter will look at:

* Convergence as change management
* Mindset, management, and newsroom culture
* Business models behind convergence
* Influence of newspaper cultures
* Audiences for news and information
* Convergence as a marketing tool
* The arrival of multi-media advertising

In August 2003 I attended a panel discussion at the annual convention of the Association for Education in Journalism and Mass Communication that discussed the business models adopted during the dot-com boom of the late 1990s. In that session in Kansas City four senior media managers talked about the lack of business models. The dot-com gold rush had a momentum of its own, and organizations appeared to have forgotten the basics of business 101. I remember asking myself at the time how a business could justify outlaying so much money on new developments without a business model? This initial section of the chapter suggests that for convergence to work, it needs some viable business models. It also needs managers to accept that convergence represents a form of research and development. Like many new ideas, convergence costs money during the initial phases of implementation.

Commercial broadcasters and newspapers in the Western world have traditionally relied on the same business model for generations: give content away in exchange for attracting eyeballs. As audiences grow, media organizations can charge more for advertising, which in turn pays for the content. Mark Choate, director of new media for the *News & Observer* in Raleigh, North Carolina, has concerns about the viability of this model. "It's my opinion that the 'free news content supported by paid advertising' model is one that isn't working and I'm not sure if it will ever work [on the web]."

The problem, he said, was the lack of a truly viable way to charge for the content. Wireless operators in countries such as Japan had other models. "Take a look at how DoCoMo is actually making money on content. There is built-in infrastructure for micro-transactions such that charging a few cents for access to a story is realistic." Choate suggested that the combination of micro-transactions and information about readers could provide a viable way to pay for local news (quoted in Outing 2000). In the previous chapter, Tribune Company president Jack Fuller said providing free information on the Internet could have been one of the newspaper industry's "biggest mistakes" because it encouraged people to expect that quality news and information would be available for free. A key issue for managers was changing public perception so that people would be willing to pay for it.

Convergence as Change Management

For managers, convergence is about dealing with change. As well as being a new way of doing journalism, convergence requires a new approach to many aspects of media organization and processes. Writer Rita Mae Brown once noted that madness was doing the same thing over and over, and expecting a different result. Publishers and broadcast executives are facing huge change as the twenty-first century evolves, and in that environment they cannot continue with business as usual. *Tampa Tribune* publisher Gil Thelen, a convergence pioneer, observed in 2000 that multi-platform publishing was a "jumbo change effort wrapped in a multi-media package" (Thelen 2000c). His opinion remained unaltered four years later. He described convergence as an extension of his work in team building and change management. In essence convergence was "a huge change process that had a multi-media ribbon on it" (Thelen 2004b). Communication was vital, he said. "You've got to say the same thing hundreds of times in dozens of venues before it reaches all levels of the organization: About the time you're getting bored hearing yourself talk, you're just beginning to really communicate effectively." Thelen also observed that ideas associated with convergence tended to run ahead of the technology available to support them (Thelen 2000c).

It is a given that convergence is happening around the world. Ifra editor Charlotte Janischewski said that the term convergence in relation to media was virtually unknown at the end of the twentieth century, but by 2002 it had become "one of the main topics of conversation in the industry [at conferences and seminars]" (2003: 34). Constantine Kamaras, CEO of Aenae Communications in Athens in Greece, told the World Association of Newspapers annual convention in Bruges in May 2002 that the web strategies of successful publishing companies had dominated the World Newspaper Congress program in 1997 and 1998, but the central theme of the 2002 congress was convergence. He issued a word of warning about costs: "The problem is that publishers have greatly exaggerated expectations of convergence as a long-term cost-saver." Kamaras warned that journalists who had to produce output in all media were spending less time doing the real reporter's job: "they are in danger of becoming content packagers rather than producers" (Kamaras 2002: 41). Kamaras launched Sportline (www.sport.gr), the largest sports portal in Greece, in 2000. Before that, he was director for management and marketing services of the World Association of Newspapers from 1997 to 2000.

Mindset, Management, and Newsroom Culture

In November 2003 Kerry Northrup reported on the progress of the Newsplex, the innovative research and training facility that had opened a year earlier at the University of South Carolina at Columbia. As of November 2003, Ifra, the international newspaper research organization based in Darmstadt, Germany, which developed the Newsplex, had detailed more than 450 convergence developments around the world. "Over the past year," Northrup said, "it's clear that convergence is no longer a fad. It is also clear that the issues are not technology." The key issues, he said, were mindset, management, and newsroom culture (Northrup 2003). Tampa's Gil Thelen described convergence as a big experiment at every step. He urged publishers considering the move to "let go of organizational inhibitions that discourage . . . risk taking" (2000c). Northrup had repeatedly pointed out that convergence was not a way to save money. It positioned a media company to expand its share of the multiple-media news and information marketplace. "Convergence is not a cost-cutting strategy. It is a growth strategy" (2002b: 15).

Thelen said unanticipated, non-traditional capital needs were common: "The need for an intranet to move text among three platforms, for digital cameras, cell phones for all for mobility, camcorders for print photographers" (2000c). Cost efficiencies could be realized through deploying editorial resources across a range of media activities rather than duplicating them for each unit, Northrup said. But it was unrealistic for a publisher to expect to go from working in one dominant medium to working in several without more people and technology. "Anyone undertaking convergence to cut expenses will be disappointed in both the financial and journalistic results" (2002a: 15). Thelen pointed out that the Tampa News Center employed more journalists in 2004 than four years earlier. "From a news point of view this has never been about reducing headcount, it's been about doing more," he said. "We have realized efficiencies in support areas such as facilities, such as HR, such as business office, and we're in the process of looking for additional efficiencies there. The idea is to find ways to support the content-generating areas—news, advertising and marketing. It's about re-allocating resources to improve our competitive situation in the market" (quoted in Glaser 2004b).

Northrup related the story of a section editor at a British newspaper who described convergence as "extra work" each morning wrestling with the computer to move yesterday's stories from the editorial system to the web. That anecdote summed up the problem, Northrup said. "There is a lot of talk and no plan. [Some] management is focused on cutting costs rather than increasing revenue." Journalists often had minimal investment in convergence because in many cases their employers were confused about what convergence really was. "Too many companies are trying to implement convergence from the inside out. They define convergence by what they already do. They start by looking at the media they own and the content those media produce. Then they look for ways to spread that content around among those various outlets and to consolidate the various editorial staffs that produce it." Northrup said media companies needed to reshape themselves from manufacturers of information products into suppliers of news and information services, in keeping with the demands of the information economy (2002a: 15). This represents yet another example of the need for a changed mindset.

Diane McFarlin, publisher of *The Sarasota Herald-Tribune*, has the appropriate mindset. Convergence was not doing more with less but "doing more with more," she told a panel discussion on journalism and business values at a conference the Poynter Institute organized in January 2002. McFarlin admitted her company was losing money from convergence but finding success through reaching a larger percentage of the target population: "There's a momentum here that I think is great for the *Herald-Tribune*." The inherent value of convergence was the ability to strengthen the individual parts. "The newspaper is now focused on depth because we're not as focused on breaking news as we originally were." But she warned about expecting to make a lot of money from convergence: "There's a suspicion that there's a formula where we've figured out how to eliminate people. We have not cut; we have continued to add and grow." McFarlin observed that convergence was misunderstood: "We generalize a great deal about convergence, when there are many flavors of convergence." At her company all three media (*The Sarasota Herald-Tribune*, SNN Channel 6, and Tribune.com) lived in one building under the same management. It was an approach, she said, that could not necessarily be emulated anywhere else (McFarlin 2002).

At the same seminar Andy Barnes, president and CEO of *The St. Petersburg Times*, acknowledged that readers were interested in getting information in different ways and agreed that organizations needed to match the interests and desires of readers to give them what they wanted. He also agreed that alliances and co-ownerships were useful for promotion—"I believe in promotion and marketing." But he worried about the cultural differences: "I'm not sure that the craft and skills are enough alike between broadcast and newspaper writing . . . or between newspaper and online. I don't think you're going to get the best newspaper report if someone is, in the first hour after the event, filing for a different medium" (Barnes 2002).

Tampa's Gil Thelen, when asked whether convergence made business sense, said his paper had quantitative and focus group results that showed the market understood and approved of their approach. "The circulation of the *Tampa Tribune* has increased in each quarter of the past two years [to March 2004]. Would I attribute that to convergence? No. We've done other things along the way. Tampa Bay Online is the now the leading portal and WFLA has maintained its rating lead. If you add it up we have increased market share. Can we claim cause and effect? I don't think so. Are we smarter and faster? On that score I would claim success" (Thelen 2004a). Early in 2004 Thelen noted a 16 percent rise in the number of people who perceived the *Tampa Tribune* as an improving newspaper compared with two years previously, based on research the company commissioned. He could not estimate how much of that rise was due to cross-promotion. "My caution is that we cannot apportion any of that to convergence activities here, but it would be equally uninformed to say that none of it is related to it," he said. Convergence appeared to have made each media segment stronger (quoted in Glaser 2004b).

Research conducted by Edgar Huang and a team from the University of South Florida reported that the readership of the Sunday edition of the *Tampa Tribune* rose 14 percent in the 12 months to the end of 2003, and the readership of the Monday to Saturday editions increased 4 percent. Over the same time the readership of *The St. Petersburg Times* declined at the same rate as the national trend: Its Sunday readership dropped 5 percent and daily readership fell 1 percent (Huang 2004: 18–19). *The St. Petersburg Times* is a high quality newspaper that has elected to concentrate on mono-media delivery. Given they are in the same market as the *Tampa Tribune* and its convergence partners, the results of their competition in a tough market would make a fascinating case study comparing the different approaches. It is worth repeating Northrup's perceptive analysis that the key issues with convergence are not technology but mindset, management, and newsroom culture. Editorial managers make decisions based on how they view the role of their media organization and its values, and what they perceive is the most appropriate model for survival and success.

Some editorial managers see convergence as simply a way to re-purpose—last century we used to say "shovel"—content from the newspaper onto a website or some other form of delivery. Hans-Dieter Baumgart, business manager of the *Rheinische Post* in Düsseldorf, Germany, said it was fallacious reasoning to expect that the public would appreciate the same content in different media. For convergence to work, content had to be of high quality and appropriate for specific media. News consumers would not tolerate long articles from the newspaper that had been dumped onto a website. Multiple use of the same content merely produced "uniform pap." It therefore was "the biggest mistake," Baumgart said, to assume that convergence would lead to lower costs because varied content needed extra people. In his opinion, convergence did not cut costs, but he maintained that if introduced properly it did not necessarily increase costs (quoted in Janischewski 2003: 35).

Business Models Behind Convergence

Why would managers at media organizations be interested in convergence? Fred Hilmer, the Harvard-educated CEO of the major Australian media group John Fairfax Holdings, suggested three scenarios to answer this question, noting that if convergence did not meet one of them it was "not going to fly." Fairfax publishes three of the four quality large-circulation newspapers in Australia, *The Sydney Morning Herald,* the *Australian Financial Review,* both based in Sydney, and *The Age* in Melbourne. The fourth quality paper is Rupert Murdoch's *The Australian,* also based in Sydney. Fairfax's online news products, smh.com.au and theage.com.au, are among the most popular online sites in Australia, along with news.com.au, the aggregated content of the nine Murdoch dailies and seven Sunday papers. Hilmer emphasized that it was not possible to provide quality journalism without the money that advertising generated. The first scenario he outlined involved potential cost sharing. Some managers believed that a "platform agnostic" approach could save money, he said. They expected lower costs through distributing and re-purposing content across several platforms. Hilmer described this concept as "way overblown," though he noted that a business case existed for convergence as a way to save some costs. But any savings would be minor. "You would not turn a company upside down for that [aspect of convergence]."

The second reason for introducing convergence involved the chance to reach as much of the audience as possible. "You keep your reader contact through convergence." Hilmer said the first law of media was that "everything fragments." Advertisers have recognized that audiences are fragmenting. John Lavine, head of the media management center at Northwestern University in Chicago, agreed that fragmentation was "the single most important trend across all media platforms" in the US. And Bob Liodice, chief executive of the US Association of National Advertisers, has gone so far as to say that "the mass media don't exist any more" (quoted in Burt and London 2004: 6). Hilmer said that if an existing business fragmented and a company could not buy new media pieces, the original company would shrink. Fragmentation made it difficult to maintain relationships with customers, Hilmer said. Media companies were subsequently fighting for a limited piece of people's time. Convergence allowed those companies to extend the relationship. "An important part of the economics of a media business is you're either a mass business or you have relationships with readers." It was important to be able to identify readers, Hilmer said. This ability was a valuable asset because it was then possible to market and to obtain valuable information from them. Good relationships with readers produced a real asset—giving something to them meant they often did not mind giving something back, such as registration details. Hilmer said *The Sydney Morning Herald* masthead meant something to readers and they had a relationship with it. "They start to use other forms of media because the first law of media is that media fragments. You keep them in the *Herald* brand through convergence. We're seeing this particularly with the Internet. One of the interesting spin-offs is that the Internet has become quite a significant channel for driving print subscriptions because people like the brand." Convergence was about "building brand equity with readers."

Fairfax's manager for corporate affairs, Bruce Wolpe, said the company had repeatedly expressed interest in strategic growth through expansion into television. "We see the potential for great benefits in terms of the production of quality news and entertainment programming by harnessing the resources of our journalism and our publications in conjunction with a television network." Contributions from Fairfax journalists and editors would greatly enhance the quality of television news in Australia. "We are clearly in a position to undertake significant initiatives in financial and business news and information on television. We believe association with our magazines and lifestyle sections can strengthen entertainment programming." Repeal of the cross-media laws that prohibited ownership of a daily newspaper and a television channel in the same city would improve the quality of television programming across the country, in both metropolitan and rural and regional areas, he said (Wolpe 2002: 4). Any restructuring of the laws would produce at least three commercial media companies evolving out of the three commercial television networks that currently operated, Wolpe said. Together with the public broadcasters, the Australian Broadcasting Corporation and the Special Broadcasting Service (the latter almost exclusively shows programs from non-English broadcasters), these companies would comprise at least "four pillars" of national media companies. They would be more diversified, of greater scale, and would have more resources to invest in programming, journalism, and content than current companies. "This will result, by any measure, in a more contestable, more competitive market with companies that are more fully capable of providing quality media services and programming" (Wolpe 2002: 5).

The third convergence scenario Hilmer described concerned operating in a defined geographic area such as a city. "Convergence is much more valuable if you're converging in Sydney rather than if you converged a Sydney newspaper with a Brisbane television station. [With this form of convergence] you get much more power in the advertising market. . . . There's a business case in terms of readers and geography." Cross promotion and marketing were also attractive by-products of convergence, Hilmer said, noting they worked best in a specific geographic area such as a city. Hilmer said his job as a manager was to assess the business and financial rationale in new ideas. "There are all these sirens on rocks, ready to seduce you with fascinating projects." Unless a new idea like convergence was viable he would not approve it. "If I don't see a rationale I don't get excited about it. That's fundamental to my job as a manager" (Hilmer 2004).

Fairfax's corporate affairs manager Bruce Wolpe said the company's major broadsheet newspapers were essentially irrelevant in every state apart from those where they were located: New South Wales (capital Sydney) and Victoria (capital Melbourne). "Were we to join with a television network, we would be instantly relevant in all states as a provider of quality media services, adding to the diversity of media services in all states." Wolpe said convergence allowed companies to exploit synergies between different media. Fairfax reporters appearing on television would bring their expertise on issues and politics and markets. "We would want our best reporters on air to provide analysis and perspective. We would want the best editors from print and broadcast to work together to affect the best coverage for each outlet" (Wolpe 2002: 7). Mark Stencel, vice president for global strategy and partnerships at washingtonpost.com, said that in an era when huge new media companies were combining the best of traditional and new media, news organizations could still benefit from focused partnerships. "It would be foolish not to try to find ways to join forces—to share costs, to take advantage of the complementary strengths and capabilities of each other's newsrooms, to help drive new users, viewers, listeners, and readers [to each other]—and, whenever possible, to make money." Stencel asserted that it was possible to do so without diluting the newspaper's editorial identity and integrity. "In fact, partnerships can help sites like ours reach users who might not otherwise have any reason to discover *Washington Post* journalism" (quoted in Tompkins 2002).

Boston-based academic Janet Kolodzy echoed the notion that convergence harnessed the benefits of online, broadcast, and print to provide news to people when and where the audience wanted it. Few people got their news from one source any more, she said. "Just look at all the ways people got news about the war in Iraq. They used TV for immediacy, online for diversity, and print for context." Editorial managers' adaptation to media fragmentation has been sluggish, and multi-platform reporting was one way to maintain links with audiences. Convergence's unfulfilled potential was connected with redistributing reporting resources, Kolodzy said. The key was to play to the strengths of each medium, and to respect those strengths. "We saw some of that in the war with Iraq," she said. "*New York Times* reporters provided updates on CNN and PBS, adding depth and nuance. We heard *Los Angeles Times* [reporters] on NPR. NBC and Fox News reporters were filing weblogs."

Managers need to understand the culture of any organization before introducing convergence. Kolodzy cited the example of *The Hartford Courant*'s collaboration with a Fox affiliate, WTIC Fox 61, to construct its own form of convergence. "Instead of trying to turn its print reporters into TV journalists, the *Courant* hired a television producer, Ellen Burns, to turn newspaper stories into TV stories. One example of this was the reporting on how New York's Cardinal Edward Egan handled priest sex-abuse cases when he was the bishop in Bridgeport, Connecticut. While the *Courant* wrote its reports, Burns packaged them for Fox 61. Instead of a weak, hurriedly produced and day-late TV story, WTIC viewers saw a well-researched report. The TV version also was aired in New York, and the *Courant* piece made it into *Newsday*, both Tribune properties. More people learned of the story in different ways." Ultimately the public was the winner, she said, because newspapers reached people who would never buy a newspaper let alone subscribe to one (Kolodzy 2003: 61).

Keijo Ketonen, chairman of one of the most successful convergence companies in Finland, the Turun Sanomat Group, said his company aimed to reach advertisers by using a variety of media. "Our principal purpose is to serve our society in the best possible way by using the communication channels most people prefer." Convergence offered major advantages for advertisers: a multi-media approach meant it was possible to reach people "around the clock" along with some interactivity and the potential of repetition to get a message across. Editor-in-chief Ari Valjakka said multi-media editorial and advertising reached consumers more efficiently, increased advertising revenues, and strengthened the *Turun Sanomat* newspaper's brand name. "It is also more appealing to younger journalists who can test themselves in print and electronic surroundings" (quoted in Skreien 2003: 47).

Sometimes it is sensible, after carefully analyzing the market, to concentrate on one skill. London is one of the most intense media markets in the world, with a wide range of national newspapers and magazines catering for a huge variety of tastes. The BBC and the commercial channels produce world class television and radio. The managing editor of *The Times*, George Brock, said his paper would be foolish trying to match the quality of the BBC by undertaking convergence, so it was more sensible to concentrate on doing excellent newspaper journalism. UK laws also strongly discouraged cross-media ownership. "Our competition comes from other newspapers. I don't mean to say we get no competition from other media, but we get much less from other media. Our principal competition is other newspapers. So that's what we specialize in doing." Brock noted the potential for convergence in regional UK cities because of different structures (Brock 2001). Specific markets aside, sometimes the reluctance to consider convergence is more a case of sticking with established patterns because they have been successful in the past. It is worthwhile here to look at research into newspaper cultures.

Influence of Newspaper Cultures

Newspaper readership in the US has continued to decline for the past half century, and part of the reason may lie in the culture of the organizations. In 1954 the combined daily circulation of newspapers in a population of 151 million was about 50 million newspapers. Half a century later the population had almost doubled to 290 million, but combined daily circulation had barely changed Monday to Friday—it had risen less than 1 percent. The Readership Institute at Northwestern University in Chicago believed an internal, organizational factor could be having an impact. The institute's hypothesis was that culture could ultimately be linked to readership, and this proved to be the case. To measure culture, the Institute used instruments developed by Robert Cooke at the University of Illinois at Chicago. More than two million employees in thousands of companies and organizations in more than 40 countries had completed the test instrument over the previous 15 years.

The Readership Institute surveyed 5,500 employees at all levels in news, advertising, circulation, and marketing at 90 newspapers. Employees completed three surveys to assess the prevalent operating culture at their newspaper and its effect on people and the business. The Readership Institute found that more than 80 percent of the newspapers studied had defensive cultures. Newspapers tended to be more defensive than other organizations that had participated in the study, and "this defensive cultural orientation seems to pervade the industry," the study found. "Even those newspapers whose cultures are the most constructive among their peers are not, when compared to results in other organizations, strongly constructive." The research also showed that newspapers with constructive cultures tended to have higher readership (Anonymous 2003). The Tribune Company president Jack Fuller acknowledged this research, saying it showed that newspapers were conservative, perfectionist, and reluctant to change—similar in culture to the military and hospitals. He noted some reasons to preserve aspects of this culture: "You want people [surgeons] to be perfectionists when they are doing delicate operations." Still, he hoped newspapers had the potential to change (Fuller 2003).

When asked how far convergence had developed around the world, Ari Valjakka replied: "Not very." Valjakka, editor-in-chief of the Turun Sanomat Group in Finland, observed that convergence was mostly at "embryo stage" because of the "traditional thinking of publishers." Newspaper consultant Dr. Juan Antonio Giner, writing in *Ideas*, the monthly magazine of the International Newspaper Marketing Association (INMA), quoted then Ifra executive director Günther Böttcher: "The problem is not the quantity of technology but the quality of management of our newspapers." Multi-media integration was open to a similar diagnosis, Giner said. "It would be a major strategic and tactical error to think that this process can be expedited simply by installing more and better technology even though that is, of course, essential" (Giner 2001b).

Charles Bierbauer, dean of the College of Mass Communications and Information Studies at the University of South Carolina, said technology could force the media to embrace convergence. "In the long term, it will have no choice. In my experience, it has always been technological innovations that have helped business develop." Bierbauer cited satellite broadcasting and text messaging in Europe as examples of innovations that changed the media. In the latter case, the public adopted this means of communication long before business took advantage of it. Bierbauer said the media industry needed to "show some imagination" with respect to convergence (Fournier 2002: 30–31).

Ari Valjakka is a member of the council that advises the Newsplex. The newspaper he edits, *Turun Sanomat* is part of one of the most converged media companies in the world. Valjakka said that keeping customers was essential for every successful business. "You get them using your services first on one channel, mostly newspapers, and strengthen the relationship and your brand at the same time by going into television, radio, the Internet [and] mobile telephones" (Valjakka 2002a, 2004b). His Danish neighbor, Ulrik Haagerup, editor of *NordJyske*, agreed: "People out there have changed their habit of media consumptions. So we have to either lose them to other media, or follow them by producing for more than one distribution channel" (Haagerup 2004). Scandinavian media groups combine innovative thinking and progressive internal cultures and are worthy of study. The Turun Sanomat Group offers an example of how investing in convergence helped expand market share and avoid excessive costs. The group is one of the case studies discussed in chapter 5, so this section looks only at the business models and revenue streams involved. The Turun Sanomat Group is based in Turku, the main city in southwest Finland. The *Turun Sanomat* (translated it means Turku News) is the third largest daily newspaper in the country.

Convergence increases the possibility for increased advertising revenues and also improves brand image. For the advertiser, convergence means reaching local customers in different but cost-effective ways. "Multi-media [advertising] allows the utilization of mixed channels combining facts in print, feelings on TV and speed in radio and net in advertising." In terms of brand recognition, Valjakka said the group reached its customers in the region at least twice each day through a multi-media approach, and it also protected the franchise. "Activity in different media keeps naturally the rivals better out of your own area." While being aware of the dangers of concentration of media, Valjakka said the positive aspect of single ownership multi-media was the opportunity for cost-effective and rationalized newsgathering. The newsroom budget was $10 million a year, of which 90 percent was allocated to the newspaper, with the rest for multi-media. Publishers would be wise avoiding the temptation to have one owner command all channels in one region: "Regulation discourages cross-ownership in free economics," he said. "In Finland this cannot be the case because there is always the state owned radio and TV, which has its operation all through the country. We know at the end that it is competition that promotes journalism" (Valjakka 2001).

The significance of localism and local content is another key component of a convergence business strategy. To use a phrase Greenpeace adopted, "think globally but act locally." Forward-thinking media executive Niklas Jonason said the great advantage and opportunity for regional newspapers should be based on their local editorial and commercial content: "The regional newspaper and its website should be the most natural place to go for relevant local information." Jonason became managing director of Citygate in Stockholm in 2002. Citygate was a joint development company that 11 Swedish newspaper publishers established in 1997, initially to coordinate their Internet efforts. By 2004 Citygate represented about 40 newspapers and had about 10 partners. The company develops websites for its owners and partners. It also has several wholly or partly owned subsidiaries, among them Internetsäljarna, a company that sells advertising on the Internet for publishers; Morningstar, a financial information site; and Reseguiden, a travel portal linked to many Swedish newspaper websites. Citygate's third business area consists of support services for owners and partners, for example, group contracts for premium cell phone services such as advertising via text messaging and streaming video or multi-media.

The challenge for regional media companies, Jonason said, was surviving in a new media world where strong brands dominated, and learning how to become multiple-media companies. "You need to make sure that there are people in each newspaper department: advertising, marketing, editorial, circulation [and] IT, who think in terms of several publishing channels." Jonason predicted that Internet Protocol television over broadband (IPTV) would introduce a new market where newspapers could match television as an advertising medium. "TV will no longer be such a well-defined medium and newspapers providing TV over the Internet will have a chance to take a bigger slice of the media pie" (quoted in Campbell 2004: 8). He encouraged major research companies such as Ifra to adopt a greater role in developing these areas.

Valjakka said it was vital for newspaper companies to explore these kinds of technological possibilities, as well as convergence. Digital technology provided a way for newspapers to diversify from their core activities into multi-media. "So the question is not whether to enter multi-media; it is a must enter situation. Most of our customers are already using electronic media. It would be hazardous not to be where our customers are." Even if the business lost money at first, it was vital to continue because convergence strengthened the newspaper's brand and in the long run kept young readers "in our sphere." Valjakka identified his group's three strengths. The first two are the highly qualified journalists and the archives and data collected from almost a century of the newspaper's existence. "The third strength is locality, which is the basis for our reporting." He said his group's niche area was localism (2001 and 2002a).

Audiences for News and Information

Publishers and broadcasters need to understand how their audiences absorb news and information. Consumers only care about what news and information they receive, not how they receive it. Convergence consultant James Gentry has often pointed out that media audiences are already converged in the way they absorb news, taking it in a variety of ways in any given day. Researcher Ramon Salaverria of the multi-media focused MMLab at Navarra University in Pamplona, Spain, said newspapers had to adapt to people's changed multiple-media consumption habits. "For young people, using alternative forms of information communication characterized by multi-media and interactivity is becoming increasingly a matter of course" (quoted in Janischewski 2003: 35). CanWest Global Communications Corporation's chief operating officer Rick Camilleri agreed that companies had to accept that the balance of power had shifted from media companies to consumers. "Consumers have limitless access to content, any time, anywhere, on demand" and his company was adapting to a "new world order" where media consumers were king (Dacruz 2004).

Ifra's Kerry Northrup noted an urgent need for market research that profiled local communities in terms of what kinds of news people used and depended on. People did not seek news and information based on the internal structures of traditional media units, he said. "Invariably consumers regard it as all one story that they are simply accessing in different ways." Successful convergence must be organized not around those individual media units but via topics and stories, and around geography or consumer needs. It was vital to put the emphasis on markets and services instead of products, Northrup said. Most convergence operations have established multi-media desks or desks with similar functions, both as ways to coordinate information flow and as symbols of the new ways of working.

Northrup suggested that the business side of media operations also needed similar desks to coordinate the financial commitment required for successful convergence (2002a: 15). He pointed out that news consumers were already leading the way in convergence. "To consumers, a story they read and watch and surf is all one story, just accessed in different ways at different times on different technology depending on what is convenient, what is required to satisfy the need to know, what fits with their media personalities." Media companies that served only part of a consumer's media mix were marginalizing themselves (Northrup 2004b). The next chapter will talk about the role of multi-media as a way to reach as much of the fragmented audience as possible.

Convergence as a Marketing Tool

In chapter 1 of this book, Northwestern University academic Rich Gordon described five models of convergence in the US. That chapter also identified the fundamental dichotomy between a business view of convergence as a tool for increased productivity and marketing, versus convergence's potential to do more thorough journalism. Gordon's first three models are attractive to business managers because they offer ways to increase the level of marketing of all platforms. Fred Hilmer of John Fairfax Holdings in Australia said convergence was a useful marketing tool, provided the marketing took place in the same geographic area. "In the one market, it is possible to save marketing costs through convergence," Hilmer said. "You get more bang for your marketing buck" (Hilmer 2004). It is useful to pause here to look at the history of media companies' involvement in new media. In some respects this history has been the story of publishers seeking to survive in an increasingly complex world. Marketing has assumed major significance in the fight for survival. In 1997 Roger Fidler published an important book about the emergence of new media and how existing media reacted to their arrival. One of the guiding principles of *Mediamorphosis*, Fidler wrote, was that in a changing environment all forms of communication media are compelled to adapt and evolve for survival. "Their only other option is to die." But new media did not arise spontaneously and independently. They emerged gradually from the metamorphosis of older media; hence the title of Fidler's book. When new forms emerged, the older media had to adapt (1997: 20).

Horst Pirker, director of Styria Media in Austria, said the main motivation for newspapers to become involved in convergence appeared to be fear. Newspapers reacted by marketing to make themselves more visible. "The actual root lies in the 'trauma of the newspaper,' that it would some day no longer be needed," he said. "Even though experience has shown that the arrival on the scene of a new medium does not make an existing medium superfluous, newspapers have a type of ingrained fear that some day the print medium will no longer exist" (quoted in Janischewski 2003: 34–35). Media managers appear not to have understood the principles of mediamorphosis, as demonstrated by their reactions to each new media in the past few generations. Newspaper managers' involvement in radio was initially defensive, in an effort to protect advertising revenue streams. The same could be said of the Internet. They had to protect the "rivers of gold" of classified and display advertising. Managers also reasoned that if newspapers did not offer news on the web, others would. They could take advantage of synergies through multiple use of content. A recipe evolved: produce news stories once and distribute them via different media. Front-end systems manufacturers helped by offering ways to automate data export from editorial systems.

The Arrival of Multi-Media Advertising

Some publishers have realized that convergence journalism requires more than shovelware. The issue, then, was how to pay for it. The next section of this chapter looks at multi-media advertising models as a potentially new form of revenue. CanWest Global's chief operating officer Rick Camilleri predicted that a new horizontal business model would expand his company's advertising base and create "huge employment opportunities." Preliminary discussions had shown advertisers were "thoroughly embracing" the concept, he said. Camilleri said shrinking newsprint costs would offset the capital investment needed for technology, but he was adamant that traditional media formats such as print newspapers would not disappear. The fate of each medium, he said, would always be dictated by the consumer (Dacruz 2004).

In January 2003 the International Newspaper Marketing Association (INMA) published a booklet designed to show that one of the most promising opportunities for publishing companies was development of a multi-media advertising strategy. This strategy involved running a campaign simultaneously across some combination of print (newspaper or magazine), broadcast (radio or traditional television or interactive television), wireless, the Internet, and some emerging platform. "This approach empowers media companies to fulfill advertisers' growing multi-media advertising needs, to respond to media fragmentation, and to address the trend of 'media multi-tasking' among consumers" (Stone 2003a: 2–3). Author Martha Stone said that the industry approach to this type of selling was in the "nascent stage." She noted that the practice of selling an advertising campaign across channels was nothing new to advertisers who used advertising agencies to place advertisements across media. "But this idea is new for media companies that want to orchestrate campaigns across their multiple media holdings" (2003a: 8). The main reasons for implementing a multi-media advertising approach included the chance to serve advertisers more effectively and efficiently, to create incremental revenue, to leverage the power of multiple media assets, and to gain a competitive edge over competitors. Commented Stone: "Cross-channel advertising strategies in media companies are practiced in earnest by only a handful of pioneers, dabbled in by dozens of companies around the world, and pondered by perhaps hundreds of companies. The new vernacular of multi-media advertising is emerging, and media companies need to understand the basic language and components involved" (2003a: 9).

Multi-media advertising took advantage of the strengths of each medium, Stone said. Each had different value when used together and separately. Television, for example, could achieve high reach in a short time, even with only one spot during a popular show. Newspapers often had the benefit of strong local ties. And the Internet reached a high percentage of people at work during the day. The synergy that occurred when media were combined in a campaign was called the "media multiplier effect," Stone said. "This is an elusive, but nonetheless powerful, component of multi-media campaigns. The phenomenon is created when cross-channel campaigns are launched in a concentrated time period, when campaign messages are cross-promoted, and when they have similar synergistic marketing messages" (2003a: 10–11).

Ifra's Newsplex organized the world's first seminar based on the concept of multi-media advertising in September 2003. The seminar was called "Adplexing: Building a cross-media advertising department [and] how to grow new revenues." The workshop was based on Stone's book, *Embracing the Power of Multi-Media Advertising*. Stone said several companies had experienced impressive incremental revenues from the launch of their cross-media advertising departments. Among the most successful pioneers were Belo in Dallas, Bell Globemedia in Toronto, Media General's Tampa cross-media operation, the giant Gannett's operations in Phoenix, and the Tribune Company's cross-media operation serving major markets in New York, Chicago, and Los Angeles (Stone 2003b: 32).

The chairman of Media General reported that sales for its multi-media flagship in Tampa totaled $13.4 million in 2001, including about $4 million in "new" money in the third quarter. The company had begun selling cross-media advertising packages two years earlier, Stone said. Tampa's integration of TBO.com, the *Tampa Tribune,* and WFLA Channel 8 reported more than $5 million in incremental revenue for 2002 (Stone 2003b: 32). Leon Levitt, executive vice president for digital media at the *Arizona Republic,* told the Adplexing seminar that cross-media advertising was "a new business model, not an initiative." Gannett offers multi-media advertising via the media it owns in the Phoenix marketplace: the *Republic* newspaper, AZCentral.com, and Channel 12. The Phoenix operation reported more than $5 million incremental revenue for 2002. "We are leveraging multi-media to gain consumer share, grow incremental profits and increase value of our media assets," Levitt said (quoted in Stone 2003b: 32). Stone reported that Gannett's Phoenix multi-media operation made about $5 million in incremental revenue with 35 to 40 clients in the first 15 months from mid-2001 (Stone 2003a: 19). The

Tribune Company's cross-media sales unit attracted between $47 million and $50 million of incremental sales in 2002 across four major markets (Chicago, Los Angeles, New York, and Hartford, Connecticut). Stone reported that the unit had enjoyed "substantial increases" in the five years to 2002 despite dismal advertising performance elsewhere in the Tribune Company and in the world economy (2003a: 19). Tribune's cross-media advertising operation in the three biggest markets in the US—New York, Los Angeles, and Chicago—totaled about $35 million in incremental revenues in 2002, and Belo reported a $10 million incremental growth for the same year (Stone 2003b: 32).

Cross-media advertising could also apply to small advertisers, Stone wrote, with several companies offering packages to advertisers with budgets of under $100,000 a year. This target group had potential for "fresh and incremental" revenues. Stone cited the example of the *Tampa Tribune,* tbo.com, and WFLA Channel 8 offering a package deal of a full page of color advertising in the newspaper, a spot on a business show on Channel 8, and an online video of the TV segment for $7,500. "The package has netted phenomenal results: dozens of advertisers have signed up, and many are impressed with the results and renew. The Media General flagship targets non-advertisers for the cross-media campaign, including dentists, funeral homes, and other small businesses" (2003b: 32).

In July 2004, MSNBC.com announced it had achieved its first profitable quarter since launching in July 1996, as part of a year of record revenue of almost $45 million. More than 80 percent of the revenue came from 60 advertisers, and more than 70 different advertisers spent more than $100,000. The company attributed the success to capturing an increasing segment of the highly desirable "at-work prime-time" audience. Dean Wright, MSNBC.com's editor-in-chief, said advertisers chose MSNBC.com because of its ability to attract a sought-after audience segment: "Hard to reach news users who wield substantial influence in their communities." Charlie Tillinghast, MSNBC.com's general manager and publisher, said the company could scarcely keep up with demand for video advertisements. "People want to come in and buy $1 million [worth of advertisements] in one shot," he said. "It's a sold-out situation for the video." MSNBC.com expected video advertisements to make up 8 percent of total revenue in the 2005 fiscal year, up from 3 percent in 2004. MSNBC.com combined the best of NBC News, MSNBC Cable, *Newsweek, The Washington Post,* CNBC, and NBC Sports. MSNBC also had content sharing agreements with Budget Travel, BusinessWeek

Online, FT.com, Forbes.com, *The Sporting News, Science* magazine, *Slate* magazine, Space.com, and BET.com (Bergman 2004).

Tobi Elkin, executive editor of *Media Daily News,* wrote in August 2004 that marketers were increasingly demanding "integrated" media plans, and "effective cross-media deals" were taking on a new importance. "Cross-media sales present a wholly different challenge on the news side, because it's hard to plan ahead and news programs don't necessarily have readily identifiable franchises," Elkin said. He noted the concerns of Mary Paris, director of business development and cross-media sales for MSNBC.com, who said that much of the challenge in cross-media sales involved educating the buyers at both online and offline agencies. "We want it to be intuitive that online reinforces on-air, and that we've got cable, online, and network properties. Our challenge with advertisers is to get them to be more integrated and to see that it really does satisfy their objectives" (quoted in Elkin 2004).

Stone noted that most advertising agencies were not structured to help media companies formulate a multi-media advertising strategy because agencies could not buy across all forms of the media. But the global advertising scene was changing rapidly, and one of the likely changes was a move toward multi-media purchasing. David Verklin, CEO of media services agency Carat North America, said he had observed more change in the advertising business in the 36 months to early 2004 than in the previous 36 years. Verklin predicted the changes in the next three years would be "even more profound." Part of the reason was the structure of the industry. Twelve international companies controlled the purchase of 95 percent of advertising time and space worldwide. Seven agency holding companies—Omnicom, Interpublic, WPP Group PLC, Publicis, Havas, Aegis and GreyGlobal—owned 90 percent of the advertising and media agencies worldwide, he said. Within network television, seven advertising agencies bought 75 percent of the inventory on those networks. And in outdoor media, three companies controlled more than half of all available space. Verklin predicted more consolidation as corporations sought to increase their size. Because of these ownership structures, Verklin also predicted a trend toward multi-media packaging of advertisements and promotions. Seven media corporations—among them News Corporation, Viacom, and Time Warner—controlled most of the media in the US, and marketers had to learn how to negotiate package deals with these giants, he said (Khan 2004).

Changes are already happening in the way advertising is allocated in the US. In June 2004 Larry Light, global chief marketing officer of McDonald's, told an advertising conference in New York that the company planned to cut its spending on prime-time television commercials even further than it had in the previous four years. "The time has come for us to agree that mass media marketing is over," Light told executives at the AdWatch conference organized by *Advertising Age* magazine. Between 2000 and 2004, McDonald's reduced its prime-time spending from two thirds of its advertising budget to one third. McDonald's spent $644 million advertising in US media in 2003. "Our brand means different things to different people." Light was acknowledging that audiences had splintered to such a degree that traditional television advertising was no longer appropriate.

Dave Burwick, chief marketing officer at beverage company PepsiCo, agreed the advertising landscape had changed. "Online will be bigger [in 2005] . . . print and outdoor will benefit from where we're going. We're taking dollars directly out of television." Carter Cast, senior vice president of marketing and merchandising at Wal-Mart, said the discount retailer's television spending would probably remain constant in 2005, but more money would be spent on "grass-roots marketing." The same month that McDonald's announced the major changes to its television strategy, TNS Media Intelligence/CMR, which tracks advertising spending trends, forecast that spending in US network TV would rise by only 9.8 percent in 2004 despite big spending on the election, compared with nearly 16 percent growth for Internet advertising (Reuters 2004).

In July 2004, consumer products giant Procter & Gamble (P&G) announced it was changing its approach to advertising because the company believed television and magazines no longer offered efficient ways to reach consumers. P&G is the second largest advertising spender in the US after General Motors, and easily the biggest spender on television advertising. In 2003, P&G allocated almost $2 billion of its advertising budget to television and another $638 million to consumer magazines. *Chicago Tribune* business writer Jim Kirk said other major Fortune 500 companies watched P&G closely because it was such a huge advertiser, often copying the company's marketing approach. Cindy Tripp, P&G's associate director of media and marketing for North America, said the company planned to consider "other communication options" such as sponsorships and product placements instead of television advertising (Kirk 2004). This could be the beginning of major change.

Stone has argued that developing a multi-media advertising initiative remained an important strategy for multi-media companies. This form of advertising benefited all players in the advertising equation, she said. "The advertiser gets a more successful campaign with greater lift from the media multiplier effect. Advertisers enjoy greater efficiency because they only have to interface with one sales team, and they receive only one bill for a multi-media campaign." Consumers appreciated the concept because it was more relevant to the way they consumed information. Stone said new methods and mindsets would be needed to build multi-media advertising departments. These included incentives for salespeople and advertisers, development of a teamwork approach between departments and between media companies and advertisers, and training programs for salespeople. Research was needed to convince salespeople and advertisers that cross-media advertising campaigns really worked. "Despite the successes and the promise of multi-media advertising, there are hurdles to overcome. Pioneers in multi-media advertising strategies have uniformly reported 'cultural' issues—that is, resistance to change from mono-media to multi-media approaches. The resisters include advertisers and salespeople." Parallels with the introduction of convergence in the newsroom should be obvious. Stone concluded that it would take time to convince people of the power of multi-media advertising strategies (2003a: 54–55).

Research studies show the need for cross-media advertising if advertisers seek to target people at specific times of the day. Television's traditional peak audience has always been between about 6 P.M. and 10 P.M. Radio attracts its biggest audiences in the early morning, part of the afternoon and during drive time. The influential Minnesota Opinion Research, Inc. (MORI) has demonstrated the attraction of the Internet during parts of the traditional working day. Digital newspapers and email gained the attention of the working population with Internet access at the office—the bulk of people have access at work—from about 9 A.M. to 11 A.M., again at lunch time, and often again in the afternoon. MORI suggested the findings could be applied to "most developed countries" after conducting research studies in the US and Sweden. "The news websites have their biggest audiences between 9 A.M. and 11 A.M., which corresponds to the prime-time of the online daily newspapers and this is what must be sold," said Rusty Coats, MORI's director for new media.

Some American online newspaper sites such as NYTimes.com or the Milwaukee (Wisconsin) Journal Interactive modified their advertisements during the day to accommodate this "online prime time." Television advertisers have understood the concept known as "day-parting" for years, but newspaper websites have only recently embraced it. Ifra editor Mari Pascual wrote that the NYTimes.com website had carried an American Airlines advertisement in the most attractive positions on its pages between 9 A.M. and 10 A.M., when the site recorded the highest levels of traffic, since 2002. Advertisers who ignored the Internet were neglecting white-collar workers with high purchasing power, she said (2004: 21).

Ben Estes, editor of chicagotribune.com, said his site reflected the day-parting concept, and was aimed at people at work from Monday to Friday. "The story people see at 9 A.M. is very different from the story at 3 P.M. that day," he said. Teresa Hanafin, editor of boston.com, said traffic for news dropped by two thirds on weekends. Boston.com attracts 110 million page views a month. "To build an audience, figure out the weekends" (Hanafin 2003). Estes said a typical week day started with shovelware at 6 A.M. but updating began soon after. The site indicated updates with timestamps and update tags. He also acknowledged the power of local news: "Our franchise is not Iraq or Washington, it's local news." His editors concentrated on utility stories: schools, weather, crime statistics, and traffic. "What you can't get from our competitors," Estes said. The site was aimed at the youth and young professional demographics. "Advertisers love that. We have those demographics in mind when we design the site each day." Hard news was featured in the mornings, whereas by noon news was more utilitarian ("mortgages, how to get a better deal on a cell phone"). In the evenings the site focused more on entertainment. The formula seemed to work. The site attracted 1.15 million page views a day, or almost 40 million a month (Estes 2003).

One of the key elements of successful convergence is supplying content that audiences want. Fairfax's Fred Hilmer noted that companies got a competitive advantage when they had skilled people producing unique content. "When people say that 'content is king' I've always understood that it meant content that had value. Junk content is not king. Content that cannot be easily replicated is king." Hilmer said that from an economic point of view proprietary content was the most attractive. "If you want to create content that cannot be easily replicated then you have to have good people" (Hilmer 2004). This increased the need to invest in training and maintaining good people. Chapter 7 discusses how to develop a smart newsroom through the use of knowledge management. The next chapter looks at the processes of creating multi-media content.

4 Convergent Journalism and Multi-Media Storytelling

A key component of convergent journalism is the ability to tell stories in different ways for multiple platforms. For some reporters accustomed to having done one form of journalism all their professional lives, this can be a daunting transition. The aim of this chapter is to develop a process for doing multi-media journalism. It offers pen portraits of multi-media journalists, and it considers the skills that a print journalist would need in the transition to multi-media storytelling. In summary, this chapter covers:

* The multi-media mindset
* Strengths and weaknesses of each medium
* Technology not the issue
* New roles for converged newsrooms
* Profiles of multi-media journalists
* Summary of multi-media skills

Many publishers around the world have been experimenting with new forms of storytelling since near the end of the last century. Newsplex training director Martha Stone said the most successful of them balanced storytelling innovation with high journalistic standards. "In the end, the journalism must continue to be credible, compelling and accurate," she said (2004: 37). This chapter proceeds from that position. Newsplex director Kerry Northrup noted that it was more difficult producing multi-media news compared with working in one medium. Reporters had to learn to work in new ways and change their habits, he said. "This is why it's so essential to train journalists to feel comfortable with the different supports at hand for news distribution and provide them with a working knowledge of the necessary technologies." The necessary culture change did not easily transfer to the newsroom. "By definition, journalists are skeptics," Northrup said. "If you tell them their work methods need innovating without explaining the reason or the extent of the adjustments, whatever steps you take to bring about these modifications stand a fair chance of failing. This is why it's necessary to know exactly where you're heading, how you plan to get there and why" (quoted in Pascual 2003b: 48). He was referring to the need for a multi-media mindset: a way of approaching new storytelling forms combined with new reporting processes.

The Multi-Media Mindset

Convergent journalism starts with the assignment editor, the person who decides what to cover and how it will be covered. This form of journalism should also be event driven, in the sense that the nature of the news event determines how many people cover that event and what media should be involved. A fire involving potential major loss of life will need a team of people to cover it adequately, in a variety of media, both because of the complexity of the situation and because of the story's visual nature. A routine news conference with the mayor may need only one reporter.

One of the advantages of convergent journalism is having several platforms available to cover a major event. When platforms are not competing but supporting each other, the most appropriate medium can be used to tell the story in the most appropriate way. Ownership convergence, discussed in the first chapter, means that media managers no longer feel the need to cover an event simply because the opposition are covering it, wasting scarce resources. Release of a think tank's report may be best covered solely by the newspaper or other print medium. The sense of partnership that this implies is the product of a changed mindset, a multi-media mindset. On other occasions, the single reporter sent to an event may need to acknowledge that the situation has developed, and be willing and able to call for back-up. This willingness to involve others also implies a change of mindset. Newsroom managers need to be flexible enough to be able to allocate resources to any developing event, based on what the reporters in the field tell them, and the managers' awareness of the power of each medium. Assignment editors need to appreciate how to take maximum advantage of each medium and to be able to decide which support is most appropriate to reporters in the field. Part of the multi-media mindset, then, involves the ability to appreciate each medium's capacity to tell a story and the respective strengths of each medium.

This approach puts a lot of responsibility on the assignment editor and the reporter in the field. Charles Bierbauer, dean of the College of Mass Communications and Information Studies at the University of South Carolina, said multi-media reporters initially needed to answer the question of what they were trying to do. This involved knowing the best way to cover an event using the range of tools, media and people available. "Convergence means defining and coordinating from the start a journalistic process adapted to a given event by making best use of the range of available tools and media." Technology was becoming increasingly easier to use, Bierbauer said, and contrary to given wisdom, journalists were less and less opposed to technology. "The new generations of professionals are steeped in digital culture that will in the future transcend all divisions: images are digital, sound is digital, text is digital and can be largely subjected to the same type of processing." But it would be foolish to suggest that this made reporting simpler, he said. "[Technology] provides a wide variety of possibilities to cover every aspect of an event." The reporter's job was to find the best possible combination of the different media. This meant extensive training and the change of mindset mentioned earlier. "Everything depends on the event being covered and the audience you seek to reach at a particular time during the day. Sometimes it is television, with the inclusion of elements from other media, that is best suited; sometimes it is a different support, print or the web, that is best suited to provide the base for journalistic processing. For the editors, this means understanding the function of each medium and bearing them all in mind in order to think of all of the possibilities that are available" (quoted in Fournier 2002: 30).

Journalists with a multi-media awareness are better equipped for dealing with a complicated, information-soaked world. Birebauer said the objective of convergence was not to reduce operating costs and transform the reporter into a one-man band. "A good writer will not necessarily be a skilled photo-reporter." But by acquiring rudimentary knowledge of audio-visual techniques, a journalist could understand the grammar and logic of each medium and extend their understanding of how best to cover an event. Birebauer said other benefits became apparent at the level of reprocessing the information. Reporters were better able to manage the information flood that was inundating modern newsrooms and to enrich that information by mobilizing resources to provide greater depth and perspective "such as supplying statistics after a sports event or the local results of a national election" (quoted in Fournier 2002: 31).

Strengths and Weaknesses of Each Medium

Each medium has strengths and weaknesses. Convergent journalism offers the opportunity to capitalize on each strength to offer a more complete form of reportage. This section of the chapter discusses each medium's strengths and weaknesses.

Print

Print is portable and flexible. Assuming a reasonable level of education, newspapers are easy to read and navigate. Print publications also offer a degree of serendipity, giving people the chance to discover things they did not expect to find. Print allows reporters to go into detail on a subject. In their book *The Myth of the Paperless Office,* Sellen and Harper noted that each medium had an "affordance." This related to the physical properties of a medium that permitted or made possible specific functions for the person perceiving or using that medium. In other words, they said, the properties of objects determined the possibilities for action. "The physical properties of paper (its being thin, light, porous, opaque, flexible, and so on) afford many different human actions, such as grasping, carrying, manipulating, folding and in combination with a marking tool, writing on. The affordances of paper, then, are about what people can do with paper" (Sellen and Harper 2002: 18). When applied to newspapers and other print publications, the affordance of paper means that people are able to navigate easily through documents, and it facilitates cross-referencing of multiple documents. Affordance also means that people can annotate documents easily, and paper's affordance allows for the interweaving of reading and writing (2002: 76).

Journalists are knowledge workers (see chapter 7 for elaboration). Sellen and Harper discovered that knowledge workers read. In their study of International Monetary Fund economists, they found the economists spent 82 percent of their time in document-related work, and more than 70 percent of this work involved reading (2002: 81). Ifra's Kerry Northrup said that print remained a preferred medium for knowledge transfer—people tended to print articles they wanted to absorb (Northrup 2004c). In terms of weaknesses, print is limited in terms of the available space. It needs to be delivered and the content is static. Once a newspaper has been distributed it cannot be updated, and at best people read news about events that happened half a day or more ago.

Television

Television's main strengths are its ability to convey emotion and a sense of real time, even if the events being described happened a week ago. It is a visual medium. Used well it can convey a lot of information in a short amount of time. People also tend to believe what they see ("seeing is believing"), so audiences tend to regard television as credible. Television is a passive medium in the sense that people sit back and receive it, and do not need to think too much—especially compared with print, which needs a certain level of literacy and which requires people to work to absorb it. This passive nature could also be seen as a weakness, in the sense that people do not concern themselves overly with the content they see. It is difficult to spend time in a room with a television set in the corner without the temptation to look at the screen. This tendency to distract people's attention is another weakness of television. The medium lacks the ability to discuss theoretical or dense material in detail, and it is not portable. The visual nature of the medium also limits news coverage. Sometimes events are not covered because of a shortage of video. Television news concentrates on events that are visual and easy to describe, like fires and car crashes. Hence the criticism that television news only covers easy or violent stories: "If it bleeds it leads."

Online Media

Online offers the immediacy of television and the ability of print to describe events in detail. It is relatively easy to use and navigate, especially since the arrival of icon-based navigation tools such as Internet Explorer and Netscape Navigator. Its main attraction is the capacity for interactivity. Ifra's Northrup described interactivity as "the gold standard" for online. "You've got to have interactivity to capitalize on what online can do." Part of online's attraction is its capacity to offer various forms of multi-media. Online also has the ability to slice and dice its audience to provide advertisers with detailed information about audiences. Chapter 2 showed that an increasing number of Americans were going online, so the medium continues to offer opportunities for advertising. But its popularity means that media companies have to invest more and more resources in updating material. People are less likely to go to sites if the multi-media or interactive materials have not been updated.

Mobile Phone Technology

Consumers tend to have a more intimate relationship with their cell phone or PDA-phone combination than any other medium. People play games on the cell phones and personalize them with individual ring tones and covers. In 2003 the ring tone industry was worth more than $2 billion. The cell phone is people's link to friends and families, and it can be used to call for help in an emergency. Cell phones offer the immediacy of broadcast and the ability to personalize information such as news alerts about specific stock prices or sport results. Their main weaknesses are short battery life, high costs for calls, and (as of early 2005) low bandwidth. Small screen sizes also limit the amount of information or video that can be seen. These limitations may improve with time.

The cell phone could develop into an ideal real-time companion for the newspaper. News on a cell phone is immediate and available in multimedia form but limited in terms of bandwidth, content and screen size. It is the exact opposite of news in a newspaper. Ifra's Kerry Northrup has suggested that in combination they "seem to cover the spectrum and share a key attribute in today's media marketplace: the convenience of portability." Journalists should view the variety of formats and media described in the previous paragraphs as a continuum, he said, and they should "build into their cross-format content a natural integration" that leads consumers from one medium to another (Northrup 2004c). This represents an application of journalism skills, not an exercise in using technology for its own sake.

Technology Not the Issue

In 2000 Ifra released a video called "Tomorrow's News," which showed a futuristic multiple-media news operation. (Ifra's website, ifra.com, offers details on how to obtain free copies of the video.) The Newsplex opened in November 2002. Newsplex director Kerry Northrup, who wrote the video script, said after watching the video or visiting Newsplex people tended to think too much about the technology on display. "But it's not about the technology." If the Newsplex produced t-shirts, the motto printed across the back in big letters would say "It's not about the technology," he said. "What is much more important than the technology is the mindset of an editorial organization that must learn to work in multiple media simultaneously and in real-time, 24/7, if it wants to be relevant and valuable to consumers in the expanding news and information marketplace." Academic Janet Kolodzy said multi-media journalism involved the ability to think differently and to be flexible in telling stories. She suggested many academics and journalists were entrenched in an old culture, compared with university students. "Our students are going to define a new [storytelling] culture." Kolodzy said she majored as a print journalist at university. When she graduated in 1978 cable television news and organizations like CNN did not exist. Multi-media storytelling involved training people "for jobs that do not exist," she said (Kolodzy 2003b).

Northrup said multi-media storytelling involved learning new skills and work flows. It necessitated better use of resources and reorganizing newsrooms to turn them into information-based operations. It was also about journalists' and managers' willingness to "preserve their ethics while adapting to the increasingly numbing pace of media change and innovation" (2002b: 17). Northrup has identified four new job descriptions for convergent journalism. He gave them the titles of "news flow manager," "story builder," the "news resourcer" and "multi-skilled reporter." Why were these roles given newly created names? "We intentionally use new words in the Newsplex," Northrup said, "because if we call it the same thing we've called it before, people tend to think the old way. New words give new ideas" (Northrup 2004c). The next section of the chapter describes them.

New Roles for Converged Newsrooms

The "news flow manager" was like an executive producer with helicopter vision who oversaw all stories in all media, ensuring that individual stories received whatever resources they needed. This manager chose the medium most appropriate for publication or distribution. The "story builder" was like a field producer who supervised individual stories, coordinating the various staff assigned to each specific story. The "news resourcer" was a specialist who provided information from archives, databases, the Internet, and other sources to all journalists and editors in a newsroom. The "resourcer" worked across stories and platforms. And the "multi-skilled reporter" gathered information and wrote stories for each of the media in the converged newsroom. Implicit in all the roles was an understanding of the strengths and weaknesses of each medium. In a nutshell, the news flow manager concentrated on the story; the story builder focused on the experience of the story; the news resourcer provided the context for the story; and the multi-media journalist provided the content, Northrup said. Each role is here discussed in detail.

News Flow Manager

The news flow manager sits at the center or hub of any process for managing news across multiple platforms. In his consulting work in newsrooms around the world, Northrup noted that the production process typically occupied the middle of the newsroom. This industrial-age system made producing a newspaper easier, but it did not improve the journalism, he said. In some cases it hampered attempts to deliver news across platforms. Northrup said modern newsrooms needed a new structure to reach audiences who got their news and information from a variety of media (Northrup 2000a). Academic Randy Covington worked as a news flow manager during Newsplex's Wireless Election Connection (see details of this coverage of primary and convention coverage in 2004 in chapter 6, and below). Covington said this person looked at the manage-

> Student journalists in the United States covered the Democratic national convention in Boston from 26 July 2004 and the Republican national convention in New York from 30 August 2004 armed only with mobile phones. It was part of a project coordinated by the Newsplex at the University of South Carolina (www.newsplex.org) to demonstrate the power of camera-equipped mobile phones as newsgathering tools. The students were creating blogs with cell or mobile phones, hence the term "moblog," which is a blend of mobile and blog.
> The project followed the success of trial coverage of the Democratic presidential primary in South Carolina on 3 February 2004. All events were partnerships between Newsplex and Cingular Wireless, one of the US mobile phone giants, called the Wireless Election Connection. Cingular provided the mobile phones. Students took photographs and emailed them with captions back to the Newsplex. Images and text were uploaded almost instantly to a mobile web log, or "moblog." Journalism professors from the University of South Carolina based at the Newsplex functioned as newsflow managers, storybuilders and news resourcers. Randy Covington, an instructor with the university's College of Mass Communications and Information Studies, worked as the newsflow manager. "This is a new form of journalism," Covington said. The project also showed how easy it was to use student journalists to provide coverage that was not always available to news orgnizations.
> "We were a lot of places the TV stations weren't. In terms of thoroughness and depth, we beat the pants off the [local] TV stations," Covington said. "For newspapers, this is a way to compete with television. For TV, it's a way to add depth to coverage." Newsplex director Kerry Northrup said this form of multi-media coverage could greatly enhance newspaper and TV coverage of events. The "moblog" was not meant as a replacement for traditional reportage, but as a way to add value to it. The "moblog" can be seen at www.RUCingular.com/election.

ment of a story "from 30,000 feet while others in the newsroom are looking at it from 10,000 feet."

In a television newsroom, the news flow manager was similar to the managing editor or executive producer. At a newspaper the role most closely resembled the managing editor or perhaps the news editor, Covington said. The big difference was that the news flow manager focused on the variety of stories, not on a specific delivery platform. "Does information a reporter has learned work best in one specific medium or should it appear across titles? What questions does that information raise and in which medium or media should they be answered? How can a graphic best be used in different delivery formats?" While others in the newsroom concentrated on individual stories, the news flow manager considered everything that was arriving and decided where it should go. "The question he or she must continuously answer is whether the news organization is feeding sufficient content to all distribution channels." Covington said it was important for the news flow manager to be located near the primary story builder and the lead news resourcer because they needed to be in constant communication. "The fact our conversations were within a cohesive management structure [during the Wireless Election Connection] made them easier and more effective" (Covington 2004b). The two editors in the "Tomorrow's News" video are news flow managers. Early in 2004 the Tampa News Center appointed one journalist in charge of sports across all media. "This is moving more toward the news flow manager position," Northrup said (2004a).

Story Builder

The story builder focuses on the experience of the story for the news consumer. The role combines the skills and talents of the television assignment editor and producer, along with the precision of the print copyeditor. University of South Carolina researcher Doug Fisher said a story builder must have an assignment editor's ability to see the various directions a story could take and a producer's ability to organize and pace a story, plus a copyeditor's eye for detail. The story builder must also be able to find resources for stories. "In the future, multi-media elements of the same story likely will flow at widely differing times and, possibly, locations," Fisher said. Instead of managing many stories for one medium, the story builder coordinated a selection of multi-media "streams" on specific topics. "While the job is much like that of a wire service editor's, it has many more elements, and it is unclear how many [stories] effectively can be done by one person," Fisher said. He worked as a story builder during the Newsplex's experimental coverage of the February 2004 Democratic presidential primary in South Carolina, where teams of students used cell phones to file text and images to a website. Each reporting team took photographs and video with a cell phone, and filed via email. (See chapter 6 for more details.) Fisher supervised, edited, and filed the stream of content produced by reporters assigned to a party that Democratic presidential aspirant Senator John Edwards held in the evening. "As story builder, I had to decide what to pursue with each type. I had to provide some direction to the reporters about topic—for instance, at one point we had no minority representatives [so] the reporters were told to seek out people of color—and I had to consider other multi-media elements" (Fisher 2004).

Northrup described a story builder as "an evolved copy or sub-editor operating at the leading edge of the news flow" who focused on development and deployment of integrated packages of story content across all appropriate media formats. "Storytelling is still the number one thing in story building, but it is but one part of the overall experience." Typically a copyeditor gets a story at the end of the news process. Northrup said part of the idea of the story builder was to move the copyediting to the front of the process. "They are quality enhancement people who choose when to introduce video or interactivity. A good story builder, like a good copyeditor, can handle several stories at a time" (Northrup 2004c).

News Resourcer

A news resourcer was a journalist who specialized in information skills. Geoff LoCicero, news resourcer at the Newsplex, said the job combined the essence of journalism—writing, editing and prioritizing news—with the best of librarianship and information management skills. The latter role involved information searching, training, infrastructure design, and content management. LoCicero said the role also involved going beyond a news organization's information to consider its knowledge. "How do you find out the valuable knowledge—about, for example, how to cover a beat [and] the beat's background—that resides in people's heads? What could these journalists share with others?" The news resourcer would also serve as a technology resource and trainer, helping colleagues track and use video and audio archives for deeper storytelling. "Journalists are often the best trainers for other journalists because they understand their needs and perspectives." LoCicero said a news resourcer could find relevant information in a timely fashion just like a good reference librarian, primarily through a news organization's archives and databases, the web and public and government records. At larger news operations, sections such as business, government and education could have their own news resourcers who specialized in those areas. "The information that a news resourcer can find will add depth and context to any report or story" (LoCicero 2004a).

In many respects the news resourcer turns information into knowledge. The relationship between knowledge management and convergence journalism is discussed in chapter 7. Northrup said this role focused on context. The aim was to provide as wide and as deep a set of references and links to the story as possible. This person needed to understand the information landscape. Northrup described the role as "an informatics journalist applying news judgment to a thorough understanding

Figure 4. Julie Nichols, Newsplex project director, confers with news resourcer Geoff LoCicero. Photo: Stephen Quinn.

of the information landscape." They were also a resource for newsroom colleagues in both technical and information areas, and a "chief editorial information officer ensuring development of key competitive assets for the news organization." Northrup emphasized that knowledge was vital for the modern newsroom, and "we ought to have systems in place to capture it." As newspapers and media organizations developed 24/7 cycles, one reporter could not handle the story alone. Reporters needed to be willing and able to share information. "So there need to be systems in place to capture knowledge [about] what worked and what did not" (Northrup 2004d).

Multi-Skilled Reporter

The multi-skilled reporter understands the strengths and weaknesses of each medium. She or he is skilled at interviewing; collecting audio, video and still images; can edit those sounds and images; and can write stories to be distributed across multiple forms of media. Convergence researcher Dr. Augie Grant said the first reporter at any breaking story needed to be able to capture as much information as possible in as many media forms as possible, and then be able to deliver the story immediately. "Freelancers have long carried this combination of skills onto the battlefield and into remote regions, sometimes being a sole witness to a story that can and should be delivered across media," he said. The multi-skilled reporter therefore had to be able to look at all the opportunities a story represented for each of the media that the journalist worked in. This reporter needed the ability to tell any story in the most appropriate format, ranging from an inverted pyramid for newspapers to a linear, broadcast narrative, to a multi-media production. "Writing across media may be one of the most difficult skills to master, but the task is made easier by the presence of story builders, editors, and others who can help refine the story for presentation," he said.

Reporters also needed to know which medium needed to be fed first. "It's not unusual for a journalist working for an organization such as CNN to feed a live television report, record a follow-up report, rewrite the script for the website, and then do a feed for a radio network, all in the space of an hour or two." Grant said it was unlikely that all journalists would be expected to master all the skills related to gathering information. "Indeed, many news directors [and] editors say they do not expect every journalist to do everything." It was more likely that in converged newsrooms journalists would possess a basic set of technical skills that meant they could handle fundamental stories. But they would also know when to call for help if the story needed more than basic skills. In collaboration with their news flow manager they would decide what backup or teams would be needed to deal with the situation. Grant said multi-skilled journalists needed to know how to operate basic newsgathering tools (chapter 6 has more details on this), but it was not necessary for them to be masters of each craft. "The primary concept of the multi-skilled journalist is not the mastery of a particular set of skills but simply the mindset that the information being gathered will be distributed through a variety of media, with recognition of the individual elements that must be captured in order to bring the story to the consumer." Grant described this as the most important role of the convergent journalist: The ability to look at a story or news event to determine what needed to be gathered and distributed for what medium (Grant 2004a).

Northrup admitted that at a regular newspaper this role tended to be the most controversial. "For the average journalist it means more work and they won't pay me any extra." But an increasing number of journalists were adopting this model, he said, and those people were very valuable and sometimes demanded premium salaries. "It is unfeasible to expect that everyone will do this. People have strengths. There will be multi-faceted journalists, but a lot of people will be specialists." All roles needed to understand the relative strengths and weaknesses of the various content formats that could be combined into the most effective approach to telling the story. "They need to be familiar with the various technologies that enable newsgathering and distribution over all four points of the convergence compass—print, video, online and mobile. In other words, you have to be multiple-media minded. The best writer in the world who refuses to have or share ideas for how to tell stories in other media is not as useful as someone with a multiple mindset" (Northrup 2004d).

Profiles of Multi-Media Journalists

Victoria Lim is one of about half a dozen multi-media reporters at the News Center in Tampa, Florida, on an editorial staff of about 400. She joined WFLA-TV in 1997 as a consumer reporter. In 1999 Lim was doing an investigation into Internet auctions, and the then news director, Dan Bradley, asked her to write a piece for the *Tampa Tribune*, along with an online version of the story. Thereafter, the *Tampa Tribune*'s features editor asked Lim to write regularly. About a year later the newspaper's senior consumer writer departed and Lim was asked to replace him to write a column on Sundays. Lim said the "biggest buzz" she got was feedback, from any medium. "It's really cool to hear feedback," she said. Lim won a Society of Professional Journalists (SPJ) award the first year she wrote her newspaper column and she won an award every year for the next three years. "That went a long way in gaining respect from my colleagues. They realized 'OK she really is a journalist and not one of those who just cares about her hair and make-up.' It took something like that to gain their respect."

Lim started writing the column before the whole staff moved into the new converged building in March 2000. This required Lim to drive across town to talk to the person who edited her column. "My editors were critical but encouraging. They told me what was wrong and they told me why." Lim said she gathered lots of information when reporting, and much of it was not used in her television packages. She often prepared multi-media packages for tbo.com, the online site, as well as the newspaper. "I personally do not care how people get [my] information, just so long as they get it. The information I provide reaches more people—people who do not watch television news, or read the newspaper—and personally I think that it makes me more marketable as a journalist" (Lim 2004).

Naka Nathaniel is a Paris-based multi-media journalist for the NYTimes.com site. He does as much work as possible while on the road, completing projects after returning to France. "I basically have to create my own desk wherever I go," he said. Because he did not need an Internet connection during the editing process he did much of that work while traveling: in the back seats of vans, waiting for flights or aboard aircraft. Nathaniel carried both Macintosh and IBM laptops, a Sony digital tape recorder, and Canon and Sony digital cameras. "I do the photo, audio, video, and Flash work on the Mac. When the feature is ready to file I use a USB thumb drive to move the elements over to the IBM. I use the PC to access the servers in New York, do email, and surf the web." Nathaniel said about 20 percent of his equipment involved peripherals such as converters, chargers, and cords, though the increased availability of wireless networks was making it easier to file. On deadline he relied on multi-media teams in the New York office for troubleshooting and polishing (quoted in Outing 2004a).

Preston Mendenhall works for MSNBC.com in London. In May 2001 he went to Afghanistan to report on the fundamentalist groups in that country, taking a tape recorder, a digital video camera, a conventional camera, a laptop and a mobile satellite telephone. With those tools he told a series of multi-media stories. Ifra writer Mari Pascual said Mendenhall had demonstrated that multi-media journalism was both possible and practicable, in terms of the potential of digital technology. The two-week trip cost MSNBC about $7,100—or about a tenth of what it would have cost to send a group of specialized journalists and their equipment (Pascual 2003a: 34).

Norwegian journalist Øyvind Woie writes for an Oslo-based newspaper, *Vårt Land*, and is keen to explore the possibilities of reporting with cell phones using multi-media messaging (MMS). Woie sent his first MMS reporting project from the CeBIT technology convention in Germany in March 2004. "Journalists must be able to produce pictures, video, audio, and web copy as well as use mobile technology and create stories and news for new broadband mobile media." Woie edited the text and posted images and text directly to the *Vårt Land* website. Each text posting was limited to 160 characters—the length of a standard Norwegian SMS message. "My thumbs hurt. They were sore and swollen after typing 32 small reports on that phone," he said. "But it was easy and as soon as it was posted I could move on. The equipment is also very light and anonymous." Norway is an ideal place to experiment with cell phone reporting. The country's national statistics office reported that 100 percent of 16-to-19-year-olds had cell phones, as did 86 percent of the overall population (Kiss 2004a).

Photojournalists have tended to be the most willing members of the newsroom to adopt technologies and embrace multi-media journalism. Keith Wheeler, deputy managing editor for multi-media and online at the *Orlando Sentinel*, said the paper employed 19 still photographers. They carried digital video cameras (Canon XL1 models) because sometimes the company wanted them to shoot video. "We treat the photographer as the expert in the field. If they cannot shoot video, they tell us." Wheeler said they worked for the *Orlando Sentinel* so their priority was to get the best still shot for the paper. "Then after they've got the still, they can shoot video. I will not ask photographers to shoot stills and video at a sporting event because that is hard to do. We probably send two people for that. But for a press conference, it's possible to set up a video camera and then let the still photographer take still shots."

It bears repeating that the role of the multi-media reporter often depends on the news event they are covering. Wheeler said multi-media journalism required several skills, including the ability to write well. "Whether it's a still picture or video you still have to put the words to it." Wheeler said technology would improve with time. "What I would really like to do is wait for the technology to change to take video and slice a still from it, once the technology is good enough." He predicted that when it was possible to take newspaper-quality stills with a video camera, some photographers would convert to that technology (Wheeler 2004b). See the NewsGear section in chapter 6 for more on this technology.

Can one person shoot video and stills at the same time? This remains one of the hottest debates: whether one person can take stills suitable for publication in a newspaper or website, and also shoot video. A key factor is usually the significance of the news event. It is possible at a minor news event, but teams should be employed for major news. As ever, news values should dictate what resources are allocated. Mark Dolan, assistant professor of visual and interactive communication at Syracuse University's school of communications, said expecting photographers to do both jobs was driven by financial concerns. "It has nothing to do with creativity or photography." He said quality would suffer. Mike Morris, program coordinator for photojournalism at the journalism and broadcast school at Western Kentucky University, was also concerned about quality. "If there's one magic moment, you can't do both," he said. "If it's a county fair, maybe you can do both."

Lou Toman, senior staff photographer for the *Sun-Sentinel* in Fort Lauderdale in Florida, learned to take video for the newspaper's website and its partner television station, Channel 4 in Miami. Toman said he routinely covered assignments with still and video cameras. "It can be done, but it's not the right way to do it. It's demanding, but I like it." Toman said it was difficult looking for the best still picture and the best video at the same time. "The right way is two photographers" (quoted in Sandeen 2000). *Orlando Sentinel* senior photographer Red Huber carried still and video cameras. Shooting still photographs continued to be his priority. "But if I also capture something on video I've still done my job, I've still touched readers or viewers with that moment. To me, using a video camera has broadened my perspective" (quoted in Gentry 1999: 9).

Mark Hinojosa, assistant managing editor for electronic news at *The Chicago Tribune*, was a former photojournalist. He said a handful of photographers at the paper carried both forms of camera. Most of the time the paper dispatched trained television photographers to shoot video. "With video, you have to include time in the equation—the event happens over time. With a still, you're looking for the moment. So your preparation and your mindset and expectations tend to be different, because you are looking for a moment, you tend to look for something that is off the event, off the action. Video tends to be on the action. The idea of going to a news event and shooting a spray of video for television is different from a still. Still photographers never stop shooting until the event is over because you never know what is going to happen. That's something word editors could never grasp." Hinojosa said reporters would never be expected to take still photographs, apart from foreign correspondents. "We give them enough training to make a decent photograph to go along with their stories." (Hinojosa 2004). Several factors ultimately determine the likelihood that one person will take still and moving pictures. Those factors include the size of the publication, the significance of the news event, the policy of each media organization, and the willingness or otherwise of individual reporters.

Summary of Multi-Media Skills

The final section of this chapter summarizes the skills journalists need to learn to be a multi-media storyteller. They must first have a multi-media mindset: the ability to think in terms of using the most appropriate tool for telling the story. Northrup related the story of an American radio station that broadcast a story in 2003 about a 95-year-old woman who enjoyed weightlifting. The radio station's website merely reproduced the transcript of the news story. Northrup suggested it would have made more storytelling sense to include a photograph on the website of the woman lifting weights. Too many media organizations continued to fall into the trap of transferring content from different media to the web unchanged, even though the Internet makes it possible to offer still or moving pictures, and allows audiences to download audio files. Static websites have been ignoring the potential of multi-media, Northrup said (Pascual 2003b: 48). Given that many print journalists are being asked to do multi-media journalism, the list of skills assumes a print background.

Converting Digital Information

All information and data must be in digital form to be moved from one medium to another. So the multi-media journalist needs to be able to capture and transfer digital data. The main tool for capturing data that are not in digital form is a scanner. These read text and images and convert them to digital files. Some examples that have been featured in the Newsplex's NewsGear suite of tools in previous years included a pen the size of a highlighter that scanned lines of documents and converted them to a text file, and a portable Visioneer XP100 scanner that read and captured images. These could be useful at news event where only paper files were provided. For moving files from one digital tool to another, the Newsplex recommends the Emergecore IT-100 network appliance. It combines wireless, Ethernet, and network connections in one appliance.

Writing for Other Media

The multi-media journalist can write for linear print publications, for the ear (radio), for the eye (television and web casts), and for interactive media such as the web (combining text and image). These are different skills and require different forms of expression. At the risk of making excuses, these are beyond the scope of this book. Consult the many excellent journalism text books to learn more about these skills. But bottom line, any journalist worth their salt knows how to write. And if you can write, and you are willing to learn, you can write for any medium. The essential talent is an ability to embrace and use words.

Interviewing

Reporters get information from a variety of sources. One of the most effective ways to collect information quickly is via interviews. The form of the interview varies depending on the medium. For the radio interview, the reporter needs to know how to ask appropriate questions to elicit useful material and to tell stories. The reporter also needs to know how to interview for television to elicit information. But because of the nature of the medium, reporters also need to know how to generate the tension or action that television sometimes needs (sometimes known as producing more heat than light). A multi-media reporter also must know how to be interviewed as an expert by partner broadcast organizations. This is called doing a "talk-back": the anchor at the broadcast outlet treats the journalist as an expert and asks questions based on the detailed report the newspaper journalist has produced or is about to write. The importance of thorough research here cannot be over emphasized.

Presenting

Preparing for a broadcast event involves the ability to get and stay relaxed despite the tension that being in an unusual environment generates. It also involves developing voice skills and learning about the most appropriate intonation and modulation that works for radio. Some voice production skills are natural; others can be learned. Anyone who does not have a speech impediment should, with adequate training, be able to write and present a radio report or voice-over track for a television package. Journalists who appear on camera will need to learn how to read from an autocue, and how to do "stand-ups" and pieces-to-camera. Print journalists also need to consider the most appropriate dress and make-up for working on the screen. One of the most crucial skills that both radio and television reporters need to learn in this age of late-breaking news is how to go live to air.

Technical Skills

Despite the availability of the "news resourcer" to help deal with technical problems, journalists will still need to possess some basic technology skills. These involve knowing how to take photographs with a digital camera and transfer those files back to base, how to edit digital audio files for the web, how to crop and size digital photographs (PhotoShop has become the default package), and how to transfer text, image and sound files to the web. It is not possible to define specific packages, apart from PhotoShop, because many different organizations use varying tools. Aspiring multi-media journalists can teach themselves some of the most basic, such as i-movie video editing software that Apple provides free with the Macintosh. Rob Curley, director of new media and convergence at the World Company, has strong opinions about learning technology like video editing. "If I were graduating right now and I did not know i-movie, I would find the dean of my j-school and punch them in the mouth because they have just done me such a disservice" (Curley 2004a).

Chapter 6 elaborates on the technology available to multi-media journalists. The next chapter introduces case studies of converged media groups around the world.

5 Case Studies of Convergence

This chapter describes convergence at places the author believes best demonstrate the process at some of the most innovative companies in the world. It looks first at one of America's convergence leaders, the World Company in the city of Lawrence in rural Kansas. Other convergence pioneers in the US are then discussed, before the chapter moves offshore to talk about convergence in Finland, one of the pioneers in western Europe. The chapter then discusses other notable integration efforts around the world. It ends with a description of the Newsplex, the hotbed of innovation at the University of South Carolina. Because of the spread of multiple-platform publishing around the world, the chapter cannot include everyone, so it focuses on places that reflect unique flavors and forms of convergence. The chapter considers:

* Convergence in Lawrence, Kansas
* Convergence in other US cities
* Convergence in Turku, Finland
* Convergence in Scandinavia
* Convergence in Southeast Asia
* Convergence in the UK
* The Newsplex and its role in convergence

In March 2004 a group of European newspaper publishers spent a week studying new media developments in the US. In those seven days the dozen executives traveled almost 3,000 miles to visit the country's main innovators. They observed the Tribune Company's cross-media advertising models in Chicago, quizzed executives of the giant Gannett group and washingtonpost.com companies in Washington, and learned about attracting advertisers in Boston (at boston.com). They also examined the New York Times Digital's successes—the company's website made $20 million profit in 2003. It is understandable that they should inspect centers of innovation at some of the biggest cities in the US. But they also visited Lawrence in Kansas, a college town of barely 80,000 souls in the middle of the country (Davies 2004: 9). In such a whirlwind tour, why would they visit such a relatively obscure place? The answer lies in the level of innovation and quality at Lawrence's World Company, one of the convergence leaders in the US and probably one of the most converged media groups in the world.

For such a small group, the World Company has received major awards. In 2003 *Editor & Publisher* and *Mediaweek* magazines named LJWorld.com, the company's main news site, as best overall newspaper site and best Internet news service in the world. Later that year the same magazines cited lawrence.com, an entertainment site launched in 2003, as the best entertainment service on the web. The Newspaper Association of America (NAA) named KUsports.com the best sports site in the US. The *Lawrence Journal-World* was the only newspaper in the country to be a national finalist in every category at the NAA's online awards. *Editor & Publisher* and *Mediaweek* then selected KUsports.com as best online sports service in the world (Roper 2003b: 14). In the summer of 2004, *Editor & Publisher* also named the Lawrence *Journal-World* as one of the best newspapers in the US in its annual list of "The 10 That Do It Right."

Such was the subsequent level of interest in the company that early in 2004 Rob Curley, the company's director of new media and convergence, scheduled a series of seminars for August that year. The aim was to restrict visitors to specific times to try to control the surging flow of media interest. The author was the last individual visitor to the site, in April 2004. The first two seminars, known as CLIK (Convergence Live in Kansas), sold out within weeks of being announced, and Curley soon organized a third to meet the demand. Major US newspaper corporations and newspapers in the top 10 US markets paid $350 a person for each two-day seminar, as did senior managers from television networks large and small, and a group of academics. Executives from newspapers in England and Canada attended along with their American counterparts. Most were especially interested in the company's convergence strategies and initiatives to reach college audiences (Outing 2004b).

Convergence in Lawrence, Kansas

The World Company has innovative leadership, one of the necessary factors Gentry identified for successful convergence (Gentry 2000). Curley said the fourth generation of the same family still owned the company, which started in Lawrence in 1881. "They all still live in Lawrence and show up at the offices each day to work. They are very progressive and forward-thinking" (Curley 2004a). The family company started one of the first cable television companies in the US, initially laying cable in 1968. It was also one of the first cable companies in the world to offer high-speed Internet access at home, and one of the first cable companies in the country to install high-speed wireless hot spots throughout a city. The company's new converged newsroom has television jack-points located throughout the building, making it easy to link digital cameras to the production studios across the road. As of mid-2004 Curley said Lawrence had the highest high-speed cable modem penetration of any city in the country. The company published one of the first daily updated websites early in 1995.

In September 2001 it merged the editorial staff of its print, web, and cable television (6News) operations, moving them into a renovated newsroom that would be the envy of much larger organizations. Curley described the newsroom as the company's "ode to convergence." The building had been the first federal building in Lawrence, completed in 1906 for the US Postal Service and enlarged in 1930. It had served as Lawrence's post office until 1965. The World Company purchased the building in 1999 and began renovating soon after. "The place looks incredible," Curley said. "I've been in newsrooms all over the world, and I've never seen one like this" (Curley 2004b).

Reporters had been doing convergence since 1999 but the new newsroom integrated their work lives, allowing journalists to talk to each other more easily. Curley's predecessor, multi-media managing editor Ann Gardner, said prior to moving into the new building journalists were "sticking [their] toes in the water" of convergence and doing some projects together. "But once we came over here [into the new building], we were kind of immersed" (quoted in Roper 2002a: 38). The company has printed *USA Today* in six states in the region since the national daily started in 1982. Despite its relatively small circulation of about 20,000 the *Lawrence Journal-World* is one of the best-looking broadsheet newspapers in the US. It has the distinction of having been redesigned twice by world famous designer Mario Garcia. LJWorld.com is the "traditional" newspaper website where news and information from the newspaper and cable station are featured. Curley's team "web-ify" content that reporters supply. They add audio-visual files and references to related stories on the web, plus links to archive and database material. Most of the video comes from the television station. Researcher James Gentry said the Flash animations, database content, and other forms of interactivity augmented print stories on the web, and helped readers "better understand the story" (Gentry 2003). The process gives the news significant depth and breadth and produces a more detailed and nuanced form of journalism.

A "command center" convergence desk dominates the center of the main floor of the newsroom. A team of people representing all three media assign, prepare and edit stories for the company's different news platforms. They listen to the police radio and coordinate stories if major news breaks. The cubicles of most reporters emanate from this desk like spokes from a central hub. A mezzanine area houses the sports and features departments. The building reflects the company's attitude to quality, Curley said. "When folks ask me what the secret has been to our websites winning so many national and international online journalism awards over the last two years, I know they haven't visited 645 New Hampshire in downtown Lawrence" (Curley 2004b). Even though the television production facilities are across the street, all of the television reporters are based in the converged newsroom. Ann Gardner admitted that one of the key issues that had to be negotiated was how to break a story. Ideally, breaking news should first be distributed on the web, because that was the fastest medium, she said. Television should be next to give people "video and a little more," and the newspaper should provide an expanded version the next morning. The danger was the potential to repeat the same story in three different media. "[That] just kills our newspaper people because if that's the case there is really no reason to pick up the paper." For the most part, Gardner said, the company's print reporters had been generous. Occasionally when a print reporter had put a lot of work into an enterprise piece the newspaper did not provide the story to the television station the night before. "In that case, we usually use a 'talk back' where the reporter will go on TV to say a few words about his story [that will appear] in the next day's paper. That way it is still his story, he gets credit for it, and the story is not completely given away—people will still want to read about it in the paper" (quoted in Roper 2002a: 40).

Curley described the form of convergence that had evolved in Lawrence by mid-2004 as "organic." Reporters knew what was expected of them, he said, and given the right environment and encouragement, convergence happened naturally. This meant that convergence had to be defined differently in Lawrence, compared with other companies. "Convergence isn't just a newspaper reporter appearing on TV or a TV reporter writing a story for print. It's anything that uses the reporter for more than one medium, regardless of the extent. And maybe it's not even reporter-specific." Convergence had also become informal, Curley said.

Figure 5. Reporters at the *Lawrence Journal-World* enjoy the airy feel of their converged newsroom. Photo: Stephen Quinn.

"But everyone knows what our expectations are—that means every reporter in the newsroom. And when they do something cool, or ask for some help to make a story work well across print, television and online, they know they're going to not only get praise from me and their direct supervisors, but that when I write my weekly report to senior management and ownership, [the reporters] know that I'm going to mention how great they did and why others in our company should be proud of them. They know this because I make sure they see a copy of the report that I send." Curley said he had taken the emphasis off print reporters appearing on television, and he had "totally killed" any unnecessary cross promotion for the sake of cross promotion. "When we mention something from one medium on another medium, it's got to be something pretty dang cool, which means we still do it three or four times a week" (Curley 2004b).

Preparation for the move to convergence was thorough. The managing editor for multi-media during the move, Ann Gardner, said executives organized a series of convergence meetings designed to involve staff. "We would just take somebody away from their boss for about four days. A group would consist of about 10 people, some from TV, the *Journal-World* and online." The aim was to let people get to know each other, she said. "Let's try to understand what the other people do, and then let's brainstorm about what we can do with this convergence thing." Afterward, newspaper people had a lot more respect for what television people did, Gardner said. "They didn't realize how much time it took to get a story on the air and get all of the components of a story together.

Plus, they just got to know each other better" (quoted in Roper 2002a: 39). This demonstrates the power of shared values and communication in any move to introduce convergence.

Researcher James Gentry has written case studies about the World Company, noting that communication was a vital component of the convergence process. Because newspaper and television reporters were initially on different computer systems and worked toward different deadlines and schedules, it was difficult coordinating story ideas and planning mutual coverage of events. (Curley pointed out that the company subsequently developed proprietary story-planning software so that all platforms were on the same system. "It's pretty cool stuff, and works incredibly well because it shows all the stories being worked on by our company and how each plans to play the story or augment it. There is even an internal blog system used daily to evaluate the paper so that editors and reporters can share thoughts on the last edition.") Gentry noted that print, television, and online editors met once a day to discuss stories and plan coverage, while reporters from television and print negotiated story ideas and coverage as often as possible. "The more the groups talk, the better the operation runs." Training was also vital, Gentry said. All reporters had specialized duties and this did not change in a converged newsroom. But reporters had to have a basic understanding of the goals, needs and demands of the other media. "Then reporters can begin developing some basic skills needed to produce stories for other media. A core group of reporters has emerged as responsible for educating and training others in their specific areas of expertise." Gentry said editors and reporters had to learn to put readers and viewers ahead of their own egos. "The important thing should be that the story is getting to the public, not which reporter or medium learned the information and then made that story an exclusive" (Gentry 2003).

Gardner said the company's small size and single-family ownership were advantages in the move to convergence. "We certainly had a smaller 'ship' to move and the fact that we have single ownership helped a lot. . . . I have a hard time imagining what it would be like if, as a lot of TV stations and newspapers are doing, maybe one person is driving back and forth between a TV station and a newspaper. I think that would be very limiting." Newspaper consultant and Newsplex training director Martha Stone said companies of any size could embrace convergence. "I think it would be just an excuse [to avoid convergence] saying that small publishers cannot do it. It's not just the big guys. There are a number of small publishers here in the US embracing the concept" (quoted in Roper 2003a: 32).

The World Company in Lawrence demonstrates that small companies can do convergence well. What lessons can aspiring converged media learn from the team in Kansas? Some themes emerged from discussion with Lawrence executives. These can be summarized as providing emphasis on local content, establishing strong links with the community, hiring quality people who produce quality content, and efficient use of technology. The next section of this chapter looks at each of these.

Local, Local, Local

Curley said his web team concentrated on doing things that said to readers "this isn't the *Lawrence Journal-World*'s website—it's your website." He believed that convergence was the future of journalism. "The problem is if the industry really starts to embrace convergence, it is not going to do it for the same reasons that this company did it. We have family ownership and it is very progressive family ownership who did this to make all our products better, and to make sure that we're around and important to our community for a long, long time. This wasn't done to increase cash flow from one quarter to the next. This was done in Lawrence because it was the right thing to do. I'm worried others will do it [convergence] to try to squeeze more profits out of a smaller staff," Curley said. Asked for the keys to sustaining community, Curley said it was important for websites to be practical. "I use [the web] all the time to find movie listings, to figure out where I'm going to eat, and to see if one of my favorite bands is playing in town this weekend. I love Lawrence.com because it helps me to live the lifestyle that I want to live" (Curley 2004a).

Different sites should have different and distinct purposes. With LJWorld.com, the aim was to tell stories that resonated with the community, using all the different tools that the Internet made possible, Curley said. "When I look at our coverage of something like a local election and see how different, interactive and in-depth it is, I'm just astonished by how lucky I am to work at a place that takes its responsibility to inform our readers so seriously" (quoted in Roper 2003b: 17). "We wanted content that would allow our audience to live their lives better." For local college students this included information on drink specials and a database of restaurants which could be searched: "Tell me which restaurant is still cooking at midnight" was an example of the kinds of content that people wanted at specific times of the day, Curley said (Curley 2004a).

Newspapers in smaller communities have a different relationship with their readers than a major market publication. In a small town things like obituaries, meeting notices, high school sports stories and family photos were every bit as important, if not more important, than the news that ended up on the front page of the newspaper, Curley said. The *Lawrence Journal-World* had a different relationship with readers than newspapers like *The New York Times* had. Small cities often have something that makes them unique. He cited places where he had previously built websites with content unique to the region: St. Augustine, Florida, was the oldest city in the nation. Hannibal, Missouri, had been the birthplace of Mark Twain. And Lawrence, Kansas, boasted the state's flagship university, one of the most tradition-rich basketball programs in the country, and a thriving local music scene. "There's not some magic recipe or template to building great local web sites. But when you focus on your community and try to bring out its flavor and character, then you're going to have something special. Building a great local web site—whether you're in a huge city or in the smallest town—is not about shoveling over the content from the paper that was printed the night before" (quoted in Roper 2003b: 14–15).

Lawrence is home to the University of Kansas, and because of the population the city has a diverse music scene. Lawrence.com has become the preferred site for music, partly because it allows people to download MP3 digital music files from local bands and find out "every possible thing that is going on in the city each day," Curley said. As of April 2004 the site's local music library of more than 1,000 songs from local bands was being downloaded as many as 14,000 times a week. Videos of local bands were also popular, Curley said, noting that the site was "100 percent database driven." This allowed for great flexibility because "no human has to get involved in doing things that computers do better, meaning our journalists can be journalists instead of content shovelers." The site also had thousands of pages of timeless information about bands and other forms of entertainment news such as film reviews, restaurant listings, and entertainment events, all linked to an email alert service (Curley 2004a). The other main site, KUsports.com, is dedicated to university sports, especially basketball and football. Coverage is extensive on days the local teams play, and fans have become faithful users of the large databases that Curley's team has designed. Curley said he loved building a huge site for a local high school sports team "that no one outside of our city could even care about. To me, that's exactly why you build it" (quoted in Roper 2003b: 15).

Coverage of the University of Kansas's basketball team, the Jayhawks, provides a compelling example of the depth and breadth of local convergence. Before the game begins, KUsports.com hosts a live online chat between readers and a *Journal-World* basketball writer. Game coverage then moves to a three-person team who provide live game updates on the website. One person writes a traditional newspaper report in AP-style that is updated after every basket. Another person updates the game brief on the home page, again after each basket. And the third person updates a detailed live box score for the game. The *Journal-World* was the first newspaper in the US to offer daily news updates via short messaging service (SMS). This technology sends text messages of about 100 characters to cell phones. Thousands of Jayhawks fans receive updates as often as four times a game throughout the season. At halftime, *Journal-World* photographers send at least one photograph back to the newspaper so that it can be posted to the online game story. Once the game is over, the online game story tells readers to get a copy of the next morning's *Journal-World* and to watch that evening's News6 broadcast. The real convergence overkill happened once the game was over, Curley said. For each game, *Journal-World* photographers publish to an online gallery 20 to 40 extra photographs that did not appear in the newspaper. Audio clips of post-game interviews with players and coaches are digitized and edited so that readers can listen to their responses. Box scores are augmented with links to detailed player pages that include game-by-game and cumulative statistics for each athlete. Video reports from sister cable station News6 are posted to KUsports.com. Because News6 does not broadcast on weekends, games played on Saturday or Sunday include special online-only post-game video reports built exclusively for KUsports.com. All post-game stories and columns are cross-linked within all of the related coverage on the website. The next morning's newspaper includes a high-profile reference telling readers what extra content is available online, and even includes fan quotes from the site's popular message boards (Curley 2004b).

Hiring Quality People to Produce Quality Content

The World Company did not embrace convergence expecting to save money through higher productivity from its staff. It adopts an almost parental attitude to its people, and this is reflected in the low turnover in the newsroom. Curley said that when the company was selecting chairs for the new converged newsroom, instead of getting expensive Aeron chairs (worth about $630 each) for just managers and much less expensive chairs for the rest of the news staff, the company chose to buy Aeron chairs for everyone. The family ownership also gives a turkey to each employee each Thanksgiving and a fruit basket at Christmas. A family member personally serves each employee at annual company holiday celebration meals in November and December. "We have hardly any job openings," Curley said. "When people come here they stay. We currently [April 2004] have our first opening in 18 months and the applications we are getting are mind-boggling, from the top journalism programs in the country, as well as from newspapers much larger than us."

Curley said other companies were doing convergence because the return on investment was right, or it meant profits for shareholders. "The World Company does it [convergence] because it's good for Lawrence ... we're not doing it to try to squeeze the staff. We probably have the largest staff of any 20,000-circulation newspaper in the world. Our television staff is similar to a market four or five times larger than what we are. Our newspaper staff fluctuates between 45 and 50 full-time journalists, editors and photographers. A normal 20,000-circulation paper would have 20 [editorial staff], maybe. All of the 45 to 50 do convergence on one level or another." As for the company's 13-person online staff, Curley said he was "amazingly lucky" to work with some of the smartest Internet developers on the planet (Curley 2004a). Small papers should not be afraid to go after what they perceived as "major-league talent," he said. The quality of a newspaper or a local website was related to the caliber of the people. "This may be a tiny newspaper in the middle of Kansas, but these folks are definitely 'driving with their brights on,' as our publisher Dolph Simons Jr. likes to say" (quoted in Roper 2003b: 15). People moved to Lawrence for the chance to do world-class journalism, Curley said. "The journalists who come here realize that world-class journalism has nothing to do with what your Sunday circulation is" (Curley 2004a). Most of the programmers who devised the websites had journalism degrees, and were aged in their twenties. "It's much easier to break the rules if you know the rules," he observed, noting that the sites' editors were all required to have traditional journalism backgrounds (Curley 2004b).

Links with the Community via Blogs

Curley said the company handed over one of the most prominent content sections of the Lawrence.com entertainment website to readers, via a series of weblogs (commonly known as "blogs"). It had become one of the most visited sections of the site. "There is a real sense of community in our blogs, and it's a community that more than likely doesn't read the daily newspaper and it probably doesn't visit our newspaper site" (quoted in Roper 2003b: 16–17). Curley said these blogs made Lawrence.com "feel and taste" like Lawrence though "maybe not the Lawrence that a

50-year-old resident knows, but definitely the Lawrence that a 20-year-old knows." The bulk of the dozen contributors as of April 2004 were volunteers seeking an outlet for their thoughts. Most were local people who enjoyed being at the center of a spirited debate. "We don't pay them and we don't edit them. However, we do read all of the blogs before they are posted so that we can check for libel" (Curley 2004a).

Some bloggers wrote several times a week, while others contributed once or twice a month. The content and writing was designed to appeal to a younger audience. Curley said the blog that first achieved success in terms of audience had been "Powder Room Confessions" by a University of Kansas student. The blog focused on her personal life, including her sex life, and often attracted dozens of comments in the feedback area. Curley said bloggers typically did not last long. Any who contributed for more than two months received free broadband Internet access at home as acknowledgment. "If a blogger on Lawrence.com lasts longer than a few months we're surprised because the churn rate is very high. It's definitely more work and more commitment than most think it will be. But those who stick with it are becoming mini celebrities in our town and are really being read." Some Lawrence.com bloggers appeared in the *Journal-World* as columnists. "And when one of our bloggers is being discussed by Rush Limbaugh, or our movie and video game bloggers are being linked to by some of the biggest niche sites on the Internet, you can really see the impact of this" (Curley 2004a).

Efficient Use of Technology

Gardner said the company was an early adopter of digital technology. Robotic cameras were installed in the newsroom in 2001 to allow reporters to do live reports (Roper 2002a: 40). Curley said the *Journal-World* was one of the first newspapers in the US to distribute content via cell phone text messages. People registered on the website to be reminded about start times for local events. At a time designated by the reader, a 100-character text message appeared on their cell phone. "We began using it for music shows and the like and now we use it for news," he said. KUsports.com users could register to have updates of scores and statistics during university football and basketball games. "It is 100 percent database driven, so no human has to get involved. It allows for great flexibility." Curley said he had been surprised by the popularity of cell phone alerts. "I also am almost always blown away by how much information the site has on it for game coverage, and no matter how much overkill we offer, our readers want it. I can't believe the number of new folks who sign up every day for us to send them SMS messages." The organization's journalists have also embraced technology. Reporters for the newspaper and television station telephone breaking news stories to an online editor. Images taken on camera phones are posted immediately on the company's websites (Curley 2004a). The World Company is one of the few fully-converged media organizations in the US. It presents an example of the "easy" end of the convergence continuum that Gentry outlined in chapter 2 (see figure 2.1).

Convergence in Other US Cities

A futuristic-looking "super-desk" or "multi-media desk" is probably the most visible feature of converged newsrooms in other parts of the US as well as in Lawrence. The best known examples are in Tampa (which operates WFLA-TV, the *Tampa Tribune*, and Tampa Bay Online, TBO.com) and the *Orlando Sentinel*, which partnered with Channel 2, an NBC affiliate, in early 2004 and produced the orlandosentinel.com website. *The Chicago Tribune* has similarly been doing its unique form of convergence since the mid-1990s. Its parent company owns WGN-TV and WGN radio, and CLTV, a 24-hour cable news network, all based in Chicago. At each site, a multi-media editor or coordinator acts as the liaison between each platform.

In Tampa, the multi-media assignment desk sits on the second floor, next to the newsrooms of WFLA-TV and tbo.com, and directly below the *Tampa Tribune* newsroom, which occupies the next floor above, with administration offices on the fourth floor. The television studios occupy the first floor. It is important to note that the multi-media assignment desk is not a central command desk, but serves in a coordinating role. Ken Knight is the multi-media coordinator. His job entails working with reporters, editors and photographers to ensure all platforms are informed about the activities, stories, information and media-related events of each other platform. "We keep no secrets here. All the platforms are kept aware of all the stories we are doing in the building. When it comes to deciding who runs what story first, and the level of exclusivity, those discussions are decided by the editors and managers. We've determined that if we try to keep secrets from the other platforms it breaks down the trust we have developed. We don't want to do that" (Knight 2004).

Media General, based in Richmond, Virginia, owns the Tampa operation. Tampa provides an example of what Northwestern University's

Rich Gordon would call ownership convergence (see chapter 1). Competition is also a feature of Tampa's media environment, supporting Fairfax CEO Fred Hilmer's assertion in chapter 3 that convergence was more likely in crowded markets focused on cities (Hilmer 2004). As of mid 2004 the Tampa Bay metropolitan area, with a population of about 3.8 million, was home to 14 television stations, 60 radio stations, seven daily newspapers, 70 non-daily newspapers, 50 magazines, and dozens of advertising companies. Many other media organizations such as Nielsen Media Research, the Home Shopping Network, and the New York Times Regional Newspaper Group headquarters were also located in Tampa.

The *Orlando Sentinel* first partnered with a cable television operation, Central Florida News 13, in 1997. The arrangement finished at the end of 2003, replaced by a partnership with an NBC affiliate, WESH Channel 2. The most obvious symbol of the paper's commitment to convergence is the multi-media desk in the middle of the newsroom that some staff have nicknamed the "bridge" in homage to the command post in the science fiction television series, *Star Trek*. As with the Tampa operation, editors from all areas of the news operation—multi-media, local news, online, photography, graphics and the day or night assignment editor—work from the 25-by-15-foot structure. It is located in the second-floor newsroom, about 50 feet from the small television studio. Keith Wheeler is associate managing editor for broadcast and online, and coordinates the production and distribution of all multi-media content. "My job, if I have to define it, is to get as much *Orlando Sentinel* content [as possible] to our media partners." Wheeler runs a staff of seven multi-media producers and six journalists devoted to online production. About 7 percent of the 350 editorial staff on the newspaper work as multi-media reporters. "I'd like all the reporters to do multi-media, but maybe 25 are regulars" (Wheeler 2004b). The Orlando operation provides an example of what Gordon would call tactical convergence, where multi-platform publishing has emerged among separately-owned companies.

The Chicago Tribune has had cameras in its newsrooms longer than most American newspapers. The Tribune Company rebuilt its Washington bureau in 1995, combining the resources of Tribune Newspapers and Tribune Broadcasting in one space. The newspaper has cameras in three suburban bureaus as well as the main newsroom on Michigan Avenue in central Chicago. *The Chicago Tribune* has between 620 and 640 editors and reporters. Mitch Locin is the paper's senior news editor for electronic news and acts as the liaison between print staff and multi-media producers. "All of this is not cheap to do. Convergence is not the idea that we can hire fewer reporters. It costs money to do."

Locin said one of his main jobs was to work with reporters, photographers and their editors to explain the reasons for doing convergence. "It takes time. There's no getting around it, it takes time out of your day. I explain things." Locin had worked as a reporter so he knew how much time some reporters wasted "kicking things around" and leaving the building for a cigarette or coffee.

Mark Hinojosa is assistant managing editor for electronic news at the paper, and Locin's boss. Hinojosa is responsible for the practicalities of convergence at the newspaper, like employing and training the organization's multi-media staff, as well as planning. He saw convergence as an opportunity to experiment with multiple distribution channels to cater for different audiences: "There is a need in the market place. Adults want news with intelligent presentation of the facts." The organization's current work flow created a newspaper. Multi-media producers created derivative products such as television or websites with that print content. "What we're trying to do is to see if we can create along the way, simultaneously for multiple platforms." The newspaper remained the primary platform because it had all the resources. "It's still the most magnificent generator of news and information. I try to imagine the newspaper as a stream. It starts with a reporter as a source of the stream, and [with time] it becomes a mighty river of a newspaper. What I want to do is to use my television and Internet producers and resources to take scoops out of that river and use them for other events" (Hinojosa 2004).

The Tribune Company could easily claim the title of America's first integrated media company, though CNN claims to be the country's oldest multi-media company. Tribune has been involved in multi-media journalism since the early 1990s. Late in 1997 the *American Journalism Review* commissioned journalist Ken Auletta to write a series of articles on "the state of the American newspaper." He chose the Tribune Company as his first subject because its peers had repeatedly nominated it as one of the most innovative media companies in the country. Then CEO John Madigan described Tribune's mantra as "synergy" and that word stuck. "Each Tribune property sees itself as an information company, not just a newspaper," Auletta reported. "Each has a multi-media desk. Each has an online newspaper. Each has a TV broadcast partner or a 24-hour cable news partner" (Auletta 1998: 30).

The Tribune Company consists of three divisions—Tribune Broadcasting, Tribune Publishing, and Tribune Interactive—and has more than 21,000 staff in the US. The company employs more than 4,500 journalists in the US and around the world. As of September 2004 Tribune Broadcasting owned and operated 26 television stations in major markets that reached more than 80 percent of US television households. The group's anchor was Superstation WGN, which could be seen in more than 57 million homes outside Chicago via cable and satellite services. Tribune Publishing owned 11 English-language daily newspapers and two Spanish-language publications, which reach about 9.3 million readers each week day and 12.5 million on Sundays. At the time Tribune newspapers had won a combined 91 Pulitzer Prizes. Tribune was the nation's second-largest newspaper publisher in terms of revenue and third in terms of total circulation. Tribune Interactive operated about 50 websites that attracted more than 10.6 million unique visitors each month. The company also owned the Chicago Cubs baseball team. Company revenues in 2003 were $5.6 billion with profits of just over $1.4 billion. One of Tribune's main business models is to make information available to consumers throughout the day: newspapers in the morning, radio during the morning and evening commute, interactive and Internet content during the work day, and television in the evening. The Tribune Company represents another example of what Gordon called ownership convergence. (See chapter 2.)

The big difference between Kansas and other converged groups was that most newspaper journalists in Kansas (45 out of about 50) worked in all forms of media, whereas in Tampa, Orlando, and Chicago each medium generally operated as a stand-alone platform, with some crossover by individual journalists. These organizations produced their own form of convergence, and most journalists tended to work for specific platforms. WFLA-TV, the *Tampa Tribune,* and TBO.com made individual news decisions about coverage but expected that sharing the same space would lead to what writer Aly Colon called "a synaptic intimacy that creates a pervasive, powerful presence" (2000). A handful of *Tampa Tribune* newspaper reporters appeared as experts on WFLA-TV, and some prepared television packages. Some WFLA-TV reporters wrote by-lined stories that appeared in the *Tampa Tribune* and on TBO.com. And TBO.com producers created news and information packages and links for viewers and readers. The WFLA logo and the *Tampa Tribune* flag both received broad recognition in the Tampa Bay marketplace because of the high levels of cross marketing. Forrest Carr, news director for WFLA-TV in Tampa, said the Tampa model of convergence was helping to create a new and stronger form of journalism. "And it's doing that with existing resources and skill sets. I am also ready to admit that our way of doing convergence is just one model" (Carr 2002a). Similar processes occurred in Orlando and with other Tribune Publishing newspapers.

A converged news operation needs processes to deal with breaking news. To deal with this *The New York Times* and *The Washington Post* have introduced the notion of the "continuous news desk." *The Los Angeles Times* calls it the "extended news desk." These desks publish breaking news online as soon as possible after stories become available, and function similarly to the rewrite desks that were common on afternoon newspapers until the 1960s. Groups of senior editorial staff at these major newspapers talk to reporters about stories the reporters are working on, or rewrite reporters' early versions of stories while they are still unfolding. Dan Bigman, associate editor of NYTimes.com, said the continuous news desk has been a catalyst for changing newspaper journalists' opinions about online, and vice versa. "The continuous news desk has changed the culture," Bigman said. When newspaper reporter John Burns wrote the breaking news story about Saddam Hussein's capture in December 2003, he was delighted to be able to publish the story on the web minutes after the event rather than waiting 12 hours before the next newspaper was published, Bigman said. Saddam Hussein was captured about 8:30 P.M. Baghdad time, which was about 12:30 P.M. Eastern time. Burns phoned the desk 30 minutes after Saddam's capture, and the story was on the web 45 minutes later.

Bigman said reporters appreciated that they were members of the New York Times Company, and not just the newspaper or the website. Joseph Russin, assistant managing editor for multi-media at *The Los Angeles Times,* said his paper created an extended news desk to get immediacy on the paper's website. "The extended news desk takes stories—wire or *LA Times* reporters' stories—and rewrites or edits the items and gets them on the website." The desk allowed the site to get ahead of stories. "We compete with *The New York Times* and *The Washington Post*. In order to be more competitive we needed to be more current." Russin said a strong push for the desk came from national and international reporters who wanted their stories published faster (Russin 2003).

The continuous news desk at *The Washington Post* is based in the newspaper's newsroom in Washington, D.C. The website, WashingtonPost.com, is located across the Potomac River in Arlington, Virginia. Robert McCartney, assistant managing editor for continuous news, said three editors and two writers worked in the continuous news department as of November 2004. They solicited and edited breaking news from beat reporters—"especially during peak web traffic hours of 9 A.M. to 5 P.M."—and also wrote their own stories. "The goal is to increase the flow of original, staff files to the web, thus distinguishing our coverage from that of others." McCartney said his team's first choice was always for the newspaper's beat reporter to write the early file for the web.

"Ideally, we want to take advantage of the beat reporter's expertise, sourcing and credibility." When a beat reporter did not have time, the option was for the beat reporter to telephone notes to the desk, where a writer produced a story under a double byline. "This arrangement encourages beat reporters to file for the web while relieving them of the burden if they're too busy." If necessary, continuous news department editors wrote stories on their own, "doing as much independent reporting as possible, and citing wires or other secondary sources" (McCartney 2004). Doug Feaver, editor of WashingtonPost.com, said stories were produced with a spirit of co-authorship between reporters and web staff. "The newspaper decides if it's a story. We decide if it leads the site. I won't put it on the site if *The Post* is uncomfortable with it," he said (quoted in Stone 2004: 36–37). Newspaper consultant Martha Stone said a by-product of the continuous news desk was a shift in approach from a mono-media to a multiple-media mindset—from thinking in terms of publishing once a day to publishing news as soon as possible (2004: 37).

The Post also partners with NBC and MSNBC affiliates, and built a television studio in its Washington, D.C., newsroom, with another in the offices of the website, washingtonpost.com, nearby in Arlington, Virginia. The studios were mostly used to allow *Post* reporters to appear on air. Mark Stencel, vice president for global strategy and partnerships at washingtonpost.com, said journalists sometimes interviewed people in the newspaper newsroom and the interview video was put on the website. *The Post* hired a group of experienced television journalists to train print journalists. They were based in the newspaper's offices and they advised television partners of appropriate content. "They can alert a partner to an upcoming story and break it by talking about it on TV or on the radio. The anchors and hosts drive users back to our website, where they can see the complete story for themselves—usually a few hours before the presses roll, when we would ordinarily post it online." Stencel said *The Post* shared its news budget every day with its partners but restricted that budget to a "very limited" group of people, noting that he "never imagined we would share as much information on a daily basis as we have" (quoted in Tompkins 2002).

Convergence in Turku, Finland

Like the World Company in Lawrence, the essence of the Turun Sanomat Group's convergence strategy in Finland is concentrating on local content. "Local, local, local" is the mantra of Ari Valjakka, editor-in-chief at

Figure 6. The continuous news desk at *The New York Times* is the nerve center for breaking news. Photo: Stephen Quinn.

the *Turun Sanomat* newspaper in Turku in the southwest region of the country. As of early 2004 the newspaper had a circulation by subscription of 120,000 and a readership of about 300,000, and covered a region of about 70 miles radius from Turku, the major city in the southwest of the country. The region had a population of about half a million. The *Turun Sanomat* newspaper was the third largest daily in the country. Valjakka emphasized that the newspaper and the written word remained the company's core business. "Focusing on words means the ability to give readers background and in-depth analysis when conveying meaning of what happens more powerfully and exactly than moving pictures. In today's turbulent world the quality newspaper's capability of explaining to people what is happening, what is behind it and where does it lead us will be the real strength of our operations."

Valjakka said a successful regional media house had three niche areas on which to build its journalistic quality. The first, as with Lawrence's World Company, was recruiting a highly qualified core group of journalists who had a through knowledge of their region and deep expertise in specific fields. "The other niche is our archives and data collected during the almost 100 years of our existence [*Turun Sanomat* first published in 1905]. The third strength is localism, which is the basis for our reporting," Valjakka said (2001). "Our approach to the question of multi-media starts with customer needs. Our niche is locality." All content was produced only in Finnish, which provided a "linguistic safeguard," Valjakka said. "Even Mr. Murdoch has avoided a country where the only recognizable words to an outsider are McDonald's, Coca Cola, and nylon" (Valjakka 2002a). Reporters did not earn extra for working in more than one medium. The enterprise agreement specified that the company could use material for any of the media outlets it chose without extra payment

(Skreien 2003: 46). Copyright was a key issue for ownership of content; to obtain the copyright on multi-media content, the company paid each journalist a one-time fee of $2,500.

All of the news content came from a newsroom staff of 160 journalists, of whom about 120 were newspaper reporters. Of those reporters, 50 also worked in television. About 100 freelancers operated out of the newsroom. Most of the editorial division staff were based in a building designed by Finnish architect Alvar Aalto (the two radio stations are the exception, housed next door). The television studios were on the ground floor of the building, and the newspaper occupied the next three floors, with advertising and administration above them. Daily meetings were held in the newspaper conference room, which kept people in shape from climbing stairs—Aalto disliked elevators and for many years they were not installed in the building. Even today the building has only two small elevators.

Valjakka said about 30 of the reporters rotated their jobs constantly, and this exposure to multi-media offered job enrichment and provided continuous on-the-job training. "Gaining competence from on-the-job training in different media naturally improves a journalist's job chances in the future and also chances for jobs in rival companies" (quoted in Skreien 2003: 45). Valjakka said he spent time negotiating with and inspiring his reporters. For example, he sometimes had to reassure journalists who concentrated on print that they were just as valuable professionals as multi-media journalists. "It is better not to try to force a journalist to do multi-media. On the contrary it takes lots of soft selling to get conservative journalists to try something new." Valjakka walked the walk as well as talked the talk, frequently appearing on television and presenting videos about the company. Before joining the Turun Sanomat Group he worked for the Finnish Broadcasting Company in Helsinki, Swedish Radio in Stockholm and the BBC External Services in London.

In 2003 the 160 journalists in the newsroom produced 9,000 broadsheet pages for the *Turun Sanomat* plus another 1,000 for other publications (12 years earlier the newspaper produced 7,500 broadsheet pages a year); 1,000 hours of local television programming (including 15-minute news programs each weekday); plus another 240 packages for MTV3 Television in Helsinki. The newsroom also produced a text news service for television and an information service for the city of Turku on cable; a 24-hour Internet news service (the oldest in Finland, which started in 1995); and news for two radio stations, Auran Aallot and Radio Majakka. Perhaps unique to Turun Sanomat, the newsroom also produced a daily telefax service. Two A4 news pages were faxed to hotels and tour operators anywhere in the world where Finnish people were on vacation. Valjakka said the service had about 30,000 readers around the world. All of the content described above came from the same editorial resources, and staff numbers had actually declined slightly even though the number of radio stations had risen from one to two. The company also produced books, with about 80 percent of content provided by their own journalists.

Digital technology was one key to success, Valjakka said. The company has always been an early adopter of new technology. The *Turun Sanomat* was the first newspaper in Europe to introduce cold metal production, in the early 1960s. And in August 2002, the newspaper published for the first time a photograph taken by a member of the public with a cell phone. The front page image was of an attempted robbery of an armored car by three men with shotguns. Valjakka said he believed it was the first picture of its kind ever taken and published by a newspaper in the world (Valjakka 2004a).

Turku Television operated out of one section of the *Turun Sanomat* newsroom. It hired its technical assistance from Visicom, a *Turun Sanomat*-owned video production company. Visicom's main business was producing company videos and sales promotion material for the business community in southwestern Finland and the rest of the country. At the same time it operated as an internal subcontractor to Turku Television, which paid for its technical services through internal invoicing. Valjakka cited an example of an interview on Turku Television that showed the power of convergence and the shared use of resources, combined with cooperation among journalists. Shipyards for building luxury liners were Turku's biggest industry. Turku Television interviewed a prominent Finnish entrepreneur, Kalle Isokallio, for Guestbook, a half-hour personality-style interview program. During the interview Isokallio claimed that shipyards were a sunset industry because of high wage costs and the stubbornness of Finnish trade unions. This was a major story. "Out of this 25-minute interview we made a piece of news for Radio Auran Aallot, wrote 100-word dispatches for the same day's Text TV, Telefax and Internet versions of *Turun Sanomat* as well as a two-minute piece of news [for] the evening Turku Television News. For next day's newspaper, we made a 300-word story for [the] economic section as well a 100-word promotion story for [the] TV page of the paper." The Guestbook program ran on Turku Television on the evening of the second day. In all, the newsroom produced eight different pieces of content for print, web, and broadcast for a total time allocation of six hours, including editing. "We see it as a very important objective that all material we produce shall be utilized as broadly as possible," Valjakka said (quoted in Skreien 2003: 44).

Valjakka said one economic advantage of multi-media was the opportunity for cost-effective newsgathering. "But you have to be aware of [the] negative sides of one-owner media in one district. Not necessarily in practice because at the end it is the individual journalist who makes the story, but mentally there is a certain element which does not belong to [a] free market information society." Valjakka said monopoly situations were unlikely in Finland because the state-owned Finnish Broadcasting Corporation, YLE, was broadcast throughout the country. To help cut costs, the *Turun Sanomat* had formed a partnership with another leading regional daily, *Aamulehti,* based about 100 miles from Turku. Because they are not competitors, each newspaper allocated three reporters to cover national events in a bureau in the capital, Helsinki. The content appeared in both newspapers, effectively doubling the bureau's output. The two dailies also shared foreign correspondents (Valjakka 2002a).

The company's basic convergence strategy was to integrate the strengths of the traditional newspaper with the power of electronic and new media. Finland's people were heavy consumers of traditional and new media. "People spend 9.1 hours a day on average with the media, more than the number of hours they spend at work. The issue is getting access to people's attention. The division of time between all possible channels is fierce." Valjakka said the company's multi-media approach meant that Turun Sanomat Media reached its customers at least twice each day in its region (population 500,000). The Turun Sanomat Group also conceived a radio and television supplement for a consortium of 14 other dailies in Finland, distributed under each paper's own flag. The *Treffi* was launched in 1995 and became the largest weekly magazine in Finland at 750,000 circulation. *Aamulehti* won the contract to produce *Treffi* for the newspaper consortium. This was possible because the circulations of the major regional newspapers did not overlap. Each week the 15 dailies shared soft news stories on food, entertainment, and travel, and this process again increased the combined output. The dailies also pooled resources for covering major international events such as the Olympics and the soccer World Cup. The company printed telephone directories, magazines, and manuals for Nokia cell phones. In 2003 the Turun Sanomat Group had an annual turnover of $230 million. About $80 million of that came from Turun Sanomat Media. "I'm glad to say that the business as a whole is profitable," Valjakka said. "The two commercial radio stations are at present the only activities still in the red. This [is] due to strong competition from about 20 other radio stations in the region as well as difficulties in selling advertising for radio" (quoted in Skreien 2003: 45–46).

The chair of the Turun Sanomat Group, Keijo Ketonen, said the company's long-term strategy was to maintain its number one position in southwestern Finland through consistent quality. "You lose your customer if you produce a lousy product. The quality of the content is at the end the most important factor for the newspaper and everything else is based on it," Ketonen said. Independent research published in 2002 showed customer satisfaction had reached 97 percent and the *Turun Sanomat* was the most quoted daily in Finland (Valjakka 2002a). Because the company already covered three in four households it was aware of a danger of market saturation, especially given that the company had decided not to expand its media operations outside a 70-mile radius. "We have to develop new ways of expanding the business in peripheral areas. This means making use of the possibilities of new media, Internet, text TV, radio, websites, and especially television. In the mass media market we believe most in [a] combination of newspaper and television" (Valjakka 2002b).

Digital technology was reducing barriers to entry, such as the traditional shortage of broadcast spectrum. Digital television freed up the broadcast spectrum, and Valjakka said it was of "utmost importance" for the newspaper business to look at the new possibilities this technology provided. "The question is not whether to enter multi-media; it is a must-enter situation." Most of his readers were already using electronic media and it would be "hazardous not to be where our customers are" even if the business was not profitable at first. "It strengthens the established brand and, in the long run, keeps young readers and customers in our sphere. Convergence for newspaper is not whether to enter, it is a must" (Valjakka 2002a).

Convergence in Scandinavia

While the Turun Sanomat Group in southwestern Finland may have achieved celebrity status because of its early embrace of convergence, neighboring countries in Scandinavia have also produced excellent integrated journalism in recent years. One of the leading converged news companies in northern Denmark is NordJyske. It started offering cross-media training for all journalists in 2002. Editor Ulrik Haagerup said only a handful of the 248 editorial employees declined the voluntary training provided by the Danish Center for Journalism Education, based in Denmark's second city, Aarhus. The newspaper's editorial staff saw themselves as "journalists" rather than emphasizing specific print or online or broadcast skills, he said, echoing the words of Rolf Lie, editor of Norway's *Aftenposten*, who said "today's journalist should say 'I'm not working in a newspaper, I'm working in news'" (Lie 2000: 1). Haagerup likened convergence to teenage sex: "Everybody thinks everybody else is doing it. The few who're doing it are not very good at it. But we brag about it anyway. The only way you can get good at it is by practice."

Some Danes had forgotten why it was so much fun to be a journalist, he said. Convergence helped them rediscover the fun. "The omnibus newspaper is dead. It only appeals to traditionalists—a small group of society." By publishing in many formats it was possible to tailor the news to many different groups in society (Haagerup 2002b). Haagerup said the training his newspaper provided was only the start of a major process, noting it was a "long road" before reporters adopted cross-media journalism as if it were second nature. "You have to make mistakes before you're good at it," he said (quoted in Stone 2004: 38–39).

Journalists at *Aftonbladet*, a daily newspaper in Sweden, experimented with hourly web radio broadcasts but these did not prove popular. Kalle Jungkvist, chief editor of *Aftonbladet* new media, said only a small number of listeners were interested, so journalists adopted a different approach and created short news broadcasts in collaboration with a Swedish radio network. Online journalists often produced interactive features independent of the newspaper's print staff, Jungkvist said. The popularity of video on the website was soaring in Sweden because of high broadband adoption. *Aftonbladet* had developed a portable television studio that one journalist could operate. Video news was available on the website, Aftonbladet.se, and journalists were experimenting with streaming live feeds using video-phones from overseas events such as the Cannes Film Festival (Stone 2004: 37).

Any attempt to include detailed case studies of convergence around the world would produce a book the size of a New York telephone directory, so this section of the chapter provides pen portraits of other interesting convergence efforts around the world. They are described in no particular order of preference.

Convergence in Southeast Asia

Convergence planning at the Ming Pao Group in Hong Kong started in 1999 and was implemented almost a year later. Paul Cheung, chief editor of *Ming Pao* newspapers, said he and his staff had several brainstorming sessions before making study tours to newsrooms at *The New York Times*, *USA Today*, the BBC, and the *Financial Times*. Senior staff spent one- and two-week attachments in those newsrooms. "We decided the integrated approach was more feasible in Hong Kong. It would prepare us for the future. The integrated newsroom is the newsroom of the future." Cooperation is a feature of Chinese culture compared with Western cultures, along with a willingness to bring in expertise. Parent company Ming Pao Enterprise Corporation hired five senior television journalists and cameramen from Hong Kong's biggest television news station, TVB, to facilitate the move to multi-media journalism. Those people became a resource in the newsroom and helped train print journalists. Cheung said competition was intense in Hong Kong. "It is not only a competition among newspapers. . . . You are competing against radio, television, etc. The time gap is so short, the investment is so huge and because things are changing so fast, those who stay ahead will come out on top" (quoted in Loh 2000: 48).

Cheung said convergence was an attractive way to offer advertising packages in a variety of media. The print edition of *Ming Pao* (circulation 100,000) was published seven days a week. *Ming Pao* had almost 300 editorial staff including 180 reporters, 30 photojournalists, 60 copyeditors, and 15 news executives. The rest included news assistants and reporters in overseas bureaus. A study by the Chinese University of Hong Kong rated *Ming Pao* the most credible Chinese-language newspaper in the special administrative region (Cheung 2001).

Maeil Business Newspaper is the main business daily in South Korea and has a daily circulation of about 900,000. The company runs a 24-hour cable television channel, Maeil Business TV News (MBN Channel 20), plus several websites. The group owns weekly publications, book publishing interests, satellite television, web-TV, and radio and wireless distribution channels for news. The company's president and publisher, Dae-Whan Chang, established the television company in 1993 and the group broadcast the first digital television program in Korea. The group's central news center disseminated knowledge and information through all possible channels. The news center chief met the chief editor of the newspaper and the news director of the TV channels to share story ideas. Chang said the group's vision was for the *Maeil Business Newspaper* to become one of the world's most prestigious "knowledge newspapers." The company had become involved in using knowledge for development in 1996 when Chang initiated *Vision Korea*, a national campaign to transform Korea into a knowledge-based economy. The newspaper campaigned to support that transformation. It organized more than 300 public events a year and sent reporters to universities and corporations to give public lectures. "We are a knowledge community," Chang said. "The only way to survive in the twenty-first century is by creating knowledge." He predicted that knowledge management would become one of the "key trends" of the twenty-first century (Chang 2001).

Convergence in the UK

National newspapers in the UK are almost exclusively mono-media (see chapter 3 for details about *The Times* in London), but some of the regional media are evolving into multi-media companies. One of the leaders is the *Manchester Evening News,* part of the Guardian Media Group. As of mid-2004 the group owned a television station in Manchester, Channel M, plus the Manchester Online website and a group of regional publications. The Guardian Media Group was a founding member of the directorate that helped fund the Newsplex. Paul Horrocks, editor-in-chief of the *Manchester Evening News,* said newspapers needed to reinvent themselves to survive, and they had to satisfy their customers' information needs.

Horrocks led by example. The *Manchester Evening News* was the first newspaper group to send journalists to the Newsplex for convergence training. Horrocks was one of the 10 journalists who spent the initial week at the Newsplex in March of 2003. Horrocks said his goal in sending people was "to give the journalists an opportunity to think about convergence and have a greater understanding of its benefits away from the atmosphere of the office." The 10 consisted of staff from all areas of the newsroom and included the news, business, sports, picture, and online editors. Horrocks described the week as an "eye-opener" for all the journalists, noting they had returned to work "eager to put into practice some of the projects they had undertaken." One of the projects involved creation of a CD, a collaboration between the web journalists of Manchester Online and the newspaper's print journalists. The CD was released on the occasion of the Manchester City Football Club's move to a new stadium. The team combined library archive reports and interviews, photographs of the old stadium, and material the club supplied. Horrocks said the CD was distributed with the newspaper and sold about 30,000 copies.

He noted that the Newsplex training had produced major changes in the way the newsroom operated. "We now have reporters who are regularly taking photographs with easy-to-use digital cameras. We bought a number of point-and-shoot digital cameras and they've now been taking them on various news assignments." Horrocks said photographers carried video cameras along with digital still cameras and there were plans to use the video cameras to obtain footage for Channel M (Veseling 2003: 102). *Manchester Evening News* journalists would be sent to other Newsplex training seminars, Horrocks said. The third group from the company spent a week at the Newsplex in April 2004.

Journalists within the Guardian Media Group also shared content between the web and newspapers, and they carried digital cameras to supply images to newspapers and websites. Writer Mari Pascual said during the war in Iraq the *Manchester Evening News* had a journalist "embedded" with British troops. Martin Dillon produced articles for the newspaper along with an online daily column (2003a: 34).

In 2001 the BBC introduced the notion of the video-journalist: one person who shoots, edits, writes, and transmits news packages. The training scheme, which began in September that year, was named the personal digital production (PDP) program. Its name during the planning stage was "BBC Rosenblum," after Michael Rosenblum, the New York-based consultant who brought the idea to the BBC. The BBC established a PDP training center at its Newcastle production office. PDP center coordinator Paul Myles said the idea was for the video-journalist to supplement traditional ways of working and to offer more "up close and personal-type" stories. He said it was inevitable that single crews would reduce the number of traditional crews but emphasized that this was not the reason for the project. "The big attraction was that this way of working would give greater access, more freedom and creativity to the video-journalist and a more honest and interesting final product."

By early September of 2004, 475 BBC staff had been trained in PDP, with the aim of training a total of about 600 by the time the project ended in March 2005. Myles said video-journalists were "making a real difference to our programs." These journalists contributed mainly to the BBC's regional news programs at 6:30 P.M., but they also filed for current affairs output, politics programming, and Welsh language and children's programs. "The range of stories and techniques are almost as numerous as the trainees themselves. Many find the access and the ability to tell stories through real people's eyes the big attraction." Myles said "multiple deployment" was another major attraction, because having extra crews gave program producers the chance to show several dimensions of a story at the same time. "For a growing number [of video-journalists], the creativity offered by the non-linear edit systems is a source of inspiration and helps producers create very individual styles" (Myles 2004). Chapter 6 has more details of the video-journalist program. Newsplex training director Martha Stone said a multi-skilling agreement between the BBC and trade unions enabled the process to start.

Mark Coyle, editorial coordinator for BBC News Interactive (nations and regions), attended a Newsplex convergence training program in 2003. There he learned how to use Visual Communicator software (made by a California company, Serious Magic). It allows journalists to tell stories by integrating multiple media into a single piece of content. Coyle returned to the BBC to train other journalists to use the technology (Stone 2004: 39).

The Newsplex and Its Role in Convergence

Innovation theorists talk about the significance of "change agents" for the successful introduction of new ideas. The Newsplex performs such a role for aspiring convergence media around the world. Founding director Kerry Northrup said the Newsplex was designed to demonstrate the possibilities of convergence. It has proved popular. Newsplex manager Julie Nichols said more than 1,500 people had toured or trained in the Newsplex in the 16 months between when it opened in November 2002 and March 2004 (Nichols 2004). Northrup said he wanted attendees to return to their newsrooms with exposure to as many of the different options as his team could show. "We definitely try to avoid being prescriptive. We don't want to tell them 'This is what you have to do.' We're trying to be much more descriptive and simply show them the many different options that are available to them, because it's hard for a newsroom to make decisions about how it wants to change when it doesn't know what the possibilities are."

Newsplex training seminars were based on the needs and wants of the groups who took part, Northrup said, but some general principles applied for all participants. "First of all, what we're training here is mindset," he said. "We're not doing production skills training—you can go lots of places for that. When people come in, the first thing we do is work with them on gaining an understanding of how the marketplace is changing [and] the new demands of news consumers." The strengths and weaknesses of different media formats were covered, and trainees discussed how to tell stories using the strengths of each medium. Groups worked through several scenarios during the one-week seminar. These scenarios covered various aspects of storytelling for multiple media "from story planning to the most effective format for distribution." Northrup said one of these scenarios always involved analyzing the media group's coverage of a specific event, and then revisiting the event from a convergence perspective, showing how journalists could cover that story via multiple media (Veseling 2003: 103).

Developing an appropriate mindset and introducing sound management practices are vital in the move to a converged newsroom. But journalists also need to know how to use the wide variety of digital tools available to them. The next chapter looks at the range of those tools, and chapter 7 considers the role of knowledge management in encouraging convergence.

Newsplex Convergence Monitor

Ifra created the Newsplex convergence monitor to track the development of convergent news-handling activities around the world. In 2003 Ifra staff sent 10,000 emails to media organizations to request their participation in the program. The emails invited companies to submit a brief profile of convergence activities to a database. Profiles can be submitted via an online interview at http://www.ifra.com/website/newsplexonl.nsf/html/index.html. The monitor provides information for executives who want to find out what others in similar situations are doing. People identified as company contacts in convergent operations will be emailed a unique web link that enables them to update their entry at any time. It is hoped the project will generate case studies that will be made available to Ifra members. For more information, contact the Ifra Newsplex at the University of South Carolina, at +1-803-737-8411 or via email at info@newsplex.org. Ifra promises it will not post online or sell any personal identifying information obtained through participation in the project. Individuals have the right to withdraw their information at any time and for any reason.

Biographies of People Mentioned in This Chapter

PAUL CHEUNG became chief editor of the *Ming Pao* daily newspaper in Hong Kong in 1995. His experience of working in television and radio early in his career helped him understand the transition from a single to a multiple-media platform. Cheung began writing for *Hua Jiau Ri Bao* in 1976 while a sophomore at the Chinese University of Hong Kong, where he majored in business management with a minor in economics. After graduating, Cheung worked as a reporter in commercial radio in Hong Kong for two years. In 1980, Cheung joined a newly launched newspaper called *Chung Bao* before venturing into television. He spent between 1982 and 1986 with the *Hong Kong Economic Journal*, the former colony's leading finance and economics publication. Cheung joined the Ming Pao Group in 1986 as chief editor of a monthly magazine and two years later became deputy chief editor of *Ming Pao*. It is the most respected Chinese-language daily in Hong Kong.

ROB CURLEY is director of new media for The World Company and the leader of World Online, the Internet division of the *Lawrence Journal-World*. He leads the company's converged operations for the combined news gathering staffs of the *Lawrence Journal-World*, the 6News television station and World Online. Curley moved to the Lawrence-based media company after a two-year stint as director of new media at *The Topeka Capital-Journal*. He previously was the manager of content development for Morris Digital Works, and had been with Morris Communications since 1996. Curley's team has developed the national newspaper website of the year—as named by either the Newspaper Association of America (NAA) or *Editor & Publisher* magazine—every year since 1998. In 2001, the NAA named Curley as the newspaper industry's "Internet pioneer of the year"—the youngest person to ever win the award. In January of 2003 the *Lawrence Journal-World* was the only newspaper in the United States to be a national finalist in every category at the NAA's online awards. The KUsports.com website was named best sports site in the nation. In 2004, the NAA named Lawrence.com the best entertainment site in the nation.

Professor JAMES K. GENTRY became dean of the William Allen White School of Journalism and Mass Communications at the University of Kansas in July 1997. Before then he was dean at the University of Nevada at Reno for five years. Dr. Gentry was on the faculty of the University of Missouri's school of journalism for almost 15 years. Under Dr. Gentry's leadership, the Kansas journalism school was singled out by its re-accreditation site-visit team for developing "pioneering curricular efforts to prepare students for media convergence" while maintaining the core emphasis on writing, reporting and critical thinking. As a consultant and facilitator, Dr. Gentry works with newspapers and other organizations to help them develop strategic focus, bring about change and improve the overall quality of management. Dr. Gentry received his Ph.D. in journalism with an emphasis on management from the University of Missouri in 1993. His current research analyzes how media organizations are changing to deal with the convergence of communications technologies. Dr. Gentry was a reporter and an editor for several papers before entering academia. In mid-2004 Professor Gentry gave up the dean's position and returned to teaching, research and consultancy.

KEN KNIGHT started as a reporter at the *Tampa Tribune* in 1988. He worked as a copy editor and page designer, and later served as a bureau chief before joining the multi-media team at the News Center in 2000. Before starting at the *Tampa Tribune*, Knight worked in radio and television news. He then went to graduate school where he concentrated on print journalism. After finishing graduate school he started at the *Tribune*. "My job entails working with reporters, editors and photographers to ensure all platforms are informed about the activities, stories, information and media-related events of each platform. We keep no secrets here. All the platforms are kept aware of all the stories we are doing in the building. When it comes to deciding who runs what story first, and the level of exclusivity, those discussions are decided by the editors and managers. We've determined that if we try to keep secrets from the other platforms it breaks down the trust we have developed. We don't want to do that."

GIL THELEN is publisher and president of *The Tampa Tribune*, a 225,000-circulation daily paper in Tampa, Florida. The *Tribune* is a national leader in multi-media journalism. Thelen joined the *Tribune* in 1998 as executive editor. He became publisher in 2003. From 1990 to 1997 he was executive editor and vice president of *The State* newspaper in Columbia, S.C. The Society of Newspaper Design in 1996 named *The State* one of America's three best midsize papers in terms of content and presentation. From 1987 to 1990 Thelen was editor and executive vice president of *The Sun News* in Myrtle Beach, S.C. In 1990 the American Society of Newspaper Editors named *The Sun News* as one of 10 examples of excellence among small newspapers. Thelen served in a number of assistant editor positions at *The Charlotte Observer* from 1978 to 1987, the last as assistant managing editor for news. He was a Washington correspondent from 1966 to 1978, first for the Associated Press, then *Consumer Reports* magazine and finally the *Chicago Daily News*. Thelen is a graduate of Duke University and did postgraduate work at Cornell University. He teaches often at colleges and universities, is an active member of the American Society of Newspaper Editors and has been a Pulitzer Prize juror. He has been chairman of ASNE's education committee, chairman of the ASNE Change Committee, served on the Journalism Values Institute and is a member of the ASNE Credibility Project. He chaired ASNE's Interactive Media Committee in 2001–2002 and chairs the Leadership Committee in 2003–04.

ARI VALJAKKA is editor-in-chief of *Turun Sanomat*, the third-largest daily newspaper and leading multi-media house in Finland. He has spent more than 40 years in journalism and communication. Half of this time he worked in radio-television and print in Finland and overseas. He has been employed with the Finnish Broadcasting Company in Helsinki; Swedish Radio in Stockholm; the BBC External Services in London; and as editor-in-chief of *Uusi Suomi* in Helsinki and *Turun Sanomat* in Turku. For the other 20 years, Valjakka has worked in the field of communication and human resources with Finnish industrial associations and on the board of a major export company, Wärtsilä, which employs 17,000 people. As a professional teacher he has been actively involved in adult education throughout his career. From 1983 to 1988 he was in charge of the senior management development programs for five international companies: Wärtsilä in Finland; Volvo in Sweden; NorskHydro in Norway; the Midland Montague Bank in the United Kingdom; and Hewlett Packard in the United States. Valjakka has published a book, *Internal Communication in the Company*, written hundreds of articles and has lectured throughout his professional career. He is a member of the International Press Institute and is on the board of the World Editor's Forum. He is president of SOS-Children Villages, a Finnish charity. Valjakka is married with one daughter, a journalist, and has two grandchildren.

KEITH WHEELER is associate managing editor for broadcast and online at the *Orlando Sentinel* in central Florida. Wheeler graduated from the University of California in Los Angeles and started a news writing career at KTLA in Los Angeles in 1991. In 1994 he moved to Palm Springs in California and worked his way up to become news director at KESQ, the ABC affiliate. He became the news director of WHPTV in Harrisburg, Pennsylvania, in 1996. He started at the *Orlando Sentinel* in September 1997 as the multi-media editor before being promoted to associate managing editor. "Convergence is not really about saving money," he said, "it's about reaching your audience in the ways they choose to receive news, because not everyone reads a newspaper, not everyone watches TV or listens to the radio."

6 Technology and Convergence

New digital newsgathering tools have the capacity to revolutionize the way that journalists find and report news. This chapter begins with examples of innovative ways that journalists in various parts of the world use digital technology to do their jobs more effectively. It then looks at the importance of technology for reporting in the twenty-first century, and describes the NewsGear, the suite of tools that Ifra selects each year for convergent newsgathering. It concludes by considering the role of the Newsplex in introducing convergence to the world's media organizations. In essence this chapter will cover:

* Innovations in newsgathering
* The arrival of the one-person band
* Reporting with cell phones
* The importance of technology for reporting
* Multi-media training at the Newsplex
* NewsGear and the Newsplex
* Digital tools and the future of the journalist

Technological determinists argue that technology drives people's decision making, and that it is the key factor in determining an organization's future. They believe, for example, that the structure of a news organization is a function of the available technology, and that technology determines news flow patterns. This approach is too simplistic. Digital tools have the potential to help journalists do their jobs much more effectively. But they are only tools. Human beings design the tools, and human beings choose how to use them. This chapter starts from that premise, while acknowledging that in the twenty-first century no journalist can survive and function effectively without a thorough knowledge of the tools available for the profession. We start by looking at some innovative ways that journalists have used tools for newsgathering.

In February 2004 a BBC political reporter in Bristol in the west of England filed what is believed to be the first television broadcast in the U.K. using a wireless Internet connection. Dave Harvey and his team filmed a report about the lack of broadband access in rural areas, edited the piece to two minutes using a laptop, and sent the report back to the studio via a wi-fi connection. After sending the piece, the team reported live from location using a digital camera, a laptop and innovative new software called Quicklink. The BBC had begun equipping journalists with Quicklink in November 2003. The software compresses video files, making transmission easier and faster. BBC reporters were the first in the world to use laptop newsgathering (LNG). The software needed a high-speed Internet connection such as a satellite phone, ISDN line, or wi-fi connection. The software delivered journalists' footage and reports using Internet protocol. Peter Mayne, executive editor of BBC Newsgathering, said the BBC had provided Quicklink to all reporters around the world. "The system was used extensively during the Iraq war by our news teams who were in the most forward positions. Being in the thick of the action [they] needed to travel with the smallest and lightest equipment possible." Mayne said the BBC could easily update the software to its correspondents. It was scaleable depending on the available Internet connection, and could operate from about 64 kilo-bits a second through to one mega-bit a second. "The greater the bandwidth, the better the picture quality," Mayne said. Video-phone connections transmitted live video at a maximum speed of 128 kilo-bits a second, while Quicklink software allowed feeds of up to one mega-bit a second depending on the speed of the connection.

Andrew Butterworth, senior engineer for the BBC's mobile technologies project, predicted an explosion in wi-fi adoption at the BBC. "We can run much tighter deadlines this way. Our team includes a cameraman, an editor and a journalist but in theory this work could be done by just one person." Butterworth said broadcast reporters enjoyed the freedom of wireless and mobile technology because it meant they did not have to wait for a satellite broadcast truck. Most wireless networks were based in cafés and public houses (bars). "At least it's a nice place to sit and edit," Butterworth said. As of early 2004 Bristol had about 50 wi-fi locations in coffee shops and bars. Commercial services charged between $1.75 and $10.50 an hour, but many community groups offered free networks.

Innovations in Newsgathering

In July 2003, 10 journalists from *The Guardian* in the UK produced part of the newspaper while sitting on the beach in Brighton, about 50 miles south of their London headquarters. Production of stories for the soft news section, called G2, for the edition of July 25 showed the potential power of wireless Internet access to free journalists from their desks. Journalists using laptops with a wi-fi (an abbreviation of wireless fidelity) card and within range of a wi-fi network can get a wireless broadband Internet connection. At the time Brighton had six wireless hotspots that allowed anyone to access a wireless broadband Internet connection for free. This anecdote demonstrated how portable journalism could be, plus the potential of technology to allow new ways of working. Deputy *Guardian* features editor Sam Wollaston said the team had an editorial meeting in a camper van on the way to Brighton. "It beats the sofa in the office," he said. The team emailed images and copy back to the office. In July 2003 a group of local enthusiasts set up a beachfront wireless connection known as the Pier to Pier project (Kiss 2003a). The availability of this new technology plus imagination and creativity made the project possible. The key was the journalists' willingness to try something new.

Also in Brighton in 2003, journalists used the city's new wi-fi networks to report directly from conferences, rather than being limited to filing from the press room. Photographers emailed images straight from the conference floor. One photojournalist described the connection as "awesome" after transferring a three mega-bit image in less than five seconds. Wi-fi cards can be installed in older personal computers, while most new laptops have the cards pre-installed. Hampshire-based wi-fi provider FreeLAN set up the service in the conference venue, the Brighton Center. FreeLAN estimated about 30 commercial wi-fi providers operated in the UK and predicted that more would appear. The center hosted the annual conferences of the Trades Union Congress and the Liberal Democrats. Costs were high, but early adopters have always tended to be willing to pay more for the opportunity to test innovative equipment. In the case of the Brighton Center, users could reserve a wi-fi connection for up to two weeks for $175 or "pay as they go" for $5 per half hour. Companies elsewhere in the UK charged journalists about $600 for a one-week connection. Wi-fi was still being developed as of late 2004, and Ethernet was still more reliable for large files. But the Brighton conferences illustrated the potential of broadband wireless Internet for liberating reporters from their desks (Kiss 2003b).

The Arrival of the One-Person Band

The availability of Quicklink and other tools ties in with a BBC plan to introduce the concept of the video journalist (VJ). In 2001 the BBC began sending groups of journalists and other volunteers to three-week training courses to learn how one person could shoot, write, edit, and package an entire news story. It was the brainchild of Michael Rosenblum, a former CBS news producer turned media consultant. Rosenblum developed this idea at NY1 in Manhattan, and the BBC gave him a contract to train about 600 BBC staff. Not all trainees were journalists; cameramen and production assistants were also permitted to apply.

The process became known as personal digital production (PDP), and the BBC established a training center in Newcastle in the north of England. Paul Myles, the PDP center coordinator, said all VJs used the firewire system to transfer footage from a camera to a laptop or desktop computer. The firewire cable linked the camera and computer, and transferred data rapidly to either. Trainees at the Newcastle center learned nonlinear editing during their three-week course. Myles said most video editing was done with Avid DV Express 3.5.4 when trainees returned to their newsrooms, but they had to learn about other software. Avid was installed on laptops and work stations in Nottingham, Norwich, Tunbridge Wells, Birmingham, Oxford, Southampton, Plymouth, Newcastle, Manchester, and Scotland. Journalists in Northern Ireland, London, and Wales used Final Cut Pro editing software on Macintosh systems (as of early 2005 Final Cut Pro was only available for the Macintosh platform). Reporters in Leeds, Hull, and Bristol were using Liquid Edition, digital video-editing software for the PC platform (Myles 2004). Chapter 4 includes more details about the project.

During the US Democratic presidential primaries in 2003, cable news network MSNBC embedded a young reporter with each of the candidates. MSNBC equipped each reporter with technology tested on the Iraqi battlefields: A small video camera, a tripod, and a powerful laptop for editing footage. The reporters were one-person operations, sending their reports to MSNBC over any available high-speed Internet connection. Often this was the nearest Starbucks coffee shop. Each reporter set up their own camera before interviewing and filming the candidate. Each edited their footage on the laptop, wrote articles for the web, and reported live. Mark Lukasiewicz, executive producer of NBC's campaign coverage called Campaign Embed, said viewers responded to the concept based on their knowledge of embedded reporters in Iraq. "The part of it we're applying to the campaigns is having reporters with the campaigns, with their stories, all the time, living and breathing it." *Wall Street Journal* reporter Michael Phillips said the network "saved a bundle" because the one-person reporting teams did not need a camera or sound crew or expensive satellite hook-ups. The candidates got "straight coverage with minimal spin" and many junior reporters had the chance to prove themselves (Phillips 2003).

Reporting with Cell Phones

Student journalists in the US covered the Democratic and Republican national conventions, in Boston in July 2004 and New York in August 2004 respectively, equipped only with cell phones. It was part of a Newsplex project to demonstrate the power of camera-equipped telephones as newsgathering tools. The project followed the success of trial coverage of a Democratic presidential primary in South Carolina on 3 February 2004 using the same technology. The Wireless Election Connection project was a partnership between the Newsplex and Cingular Wireless, one of the US cell phone giants. Cingular provided Nokia 3600 cell phones to students who used them to capture photographs of the primary and related events. Working in pairs, the students emailed one or two paragraphs of caption to the Newsplex office along with photographs. Images and text were uploaded almost instantly to a mobile weblog, known as a "moblog." These were blogs designed to work with cell phones. Postings were created by sending a multi-media message or email from the phone to a website linked to the blog. The email's subject line became the headline for the posting, and the message text the body of the story. Software inserted any attached photograph in the posting as a thumbnail linked to a full-size image.

While students were in the field, journalism professors from the University of South Carolina based at the Newsplex and at the conventions edited the information, chose where to place stories on the site and added extra information. They performed the roles outlined in chapter 4 of "news flow manager," the "story builder," and "news resourcer," while students acted as "multi-skilled reporters." Randy Covington, a professor with the university's College of Mass Communications and Information Studies, oversaw the project as the news flow manager. "This is a new form of journalism," Covington said. "In terms of thoroughness and depth, we beat the pants off the [local] TV stations," Covington said. "For newspapers, this is a way to compete with television. For TV, it's a way to add depth to coverage. This Wireless Election Coverage is an example of how technology and resourcefulness are blurring the lines between traditional print and broadcast journalism in a way that is creating new forms of storytelling."

In mid-2004 Ifra promoted Northrup to director of publications, and Harald Ritter replaced him as Newsplex director at Ifra's German headquarters. Covington became Newsplex director at the South Carolina site. Julie Nichols remained the Newsplex projects director and Geoff LoCicero the Newsplex news resourcer. Northrup said the "moblog" was not meant as a replacement for traditional reportage, but as a way to add value to it. Most participants agreed that the biggest lesson was in the redefinition of the concept of story. "The moblogging technique enabled the telling of a big story through hundreds of small stories, often stories so small that they were below the threshold of attention in traditional newspaper and television minds. It is pointillism applied to publishing." Northrup suggested that advanced mobile newsgathering technologies would force news organization to rewrite their business strategies and editorial work flows, and recast themselves as information service companies. They would have to deliver the information that customers wanted in whatever format and whenever and wherever they wanted it (Northrup 2004a: 18).

Covington acted as newsflow manager at the Democratic national convention in Boston in July 2004. The Democratic Party accredited 15,000 media representatives, Covington said, and they included some of the smartest, most aggressive and insightful journalists in the world. But he was disappointed with the coverage. "We should have been able to sit back and see journalism at its best, as this army of reporters covered one of the closest and most hard-fought presidential contests in our nation's history. But as I sat in my Boston hotel room and walked around the convention itself, I was struck by how little journalism I was seeing." Covington said he saw the media repeatedly covering the same handful of stories in a formulaic way. "More importantly, they were covering those stories to the exclusion of so much else."

This formula of "flogging the big story until the life is drained out of it" worked well with big stories like the war in Iraq. "But it doesn't work near as well when the big story is Teresa Heinz Kerry's suggestion to a print journalist to 'shove it,'" Covington said. He took a team of eight student journalists to Boston, where they used camera-equipped cell phones to file to a mobile weblog, or moblog. "It is a new format [and] a new form of journalism. The rules are still being written, which is why I find it so satisfying." Because they were not shackled by the competitive pressures and demands of old media, new media such as moblogs were free to try new things and cover a broader range of stories, Covington said. "It's not that new media are any better than old media. But rather, new media are free to do so much more." Media usage trends supported the conclusion that old media would continue to shed viewers and readers as they narrowed their focus and "dumbed down" their coverage. This would open the door for newer styles and formats to attract niche audiences, which would continue to grow, he said. "It may be a bit of an overstatement, but I believe new media do indeed represent new hope for journalism" (Covington 2004b).

Geoff LoCicero, the Newsplex news resourcer, supported the students in the field in Boston from the Newsplex in Columbia, South Carolina. See chapter 4 for details of the role of the news resourcer. One challenge was finding relevant links and background information to supplement the moblog site. "While it's easy enough to find related sources in advance, it's often tougher to find information online to complement or add depth to a single posting from the field." This role represented a key aspect of the news resourcer position. Important links gave a moblog posting added depth and meaning, LoCicero said. "An example would be adding links about John Kerry's views on health and education to a post quoting a voter who supports Kerry's health and education platforms. Rather than just quoting a voter's praise of Kerry's ideas on education, a link to Kerry's platform gives the reader more depth and information." Another example would be linking to demographic data for a particular voting precinct, he said. One effective strategy for providing complementary news stories was to use databases such as Lexis-Nexis. "The advanced searching features of these databases are far better than any search engine. Any hit in the database provides detailed information about the publication, author and date, which makes it easier to find the story online at a particular news site so that it can be linked." Moblogging had become a powerful tool for training and projects (LoCicero 2004b).

In Austria, photojournalists at the *Vorarlberger Nachrichten* in Schwarzach have carried camera-enabled cell phones as well as standard digital cameras since 2002. On arrival at a news event, they were all expected to take photographs with the cell phone and immediately send images to the newspaper's website before using their other cameras. Jochen Hofer, editor-in-chief of Vorarlberger Online, said the policy started in July 2002 after a photographer found he could not send images taken with his digital camera via the local phone lines because the files were too large and the connection poor. "That's why we tried our MMS [multi-media messaging system] mobiles. We knew we had to win time,"

Hofer said. MMS images were sent to online editors via the cell phone and the photojournalist also telephoned the office to dictate two or three sentences about what happened. Hofer said his photojournalists used MMS for most news events such as accidents, fires, and avalanches, plus sporting events. "The photographers get a new mobile every year, so the quality of the pictures sent by MMS is improving steadily" (quoted in Northrup 2004a: 18).

The Importance of Technology for Reporting

These examples and anecdotes illustrate the exciting range of digital tools available to journalists. Digital technologies can help make an effective news organization even more productive, allowing them to produce appealing multiple-media news stories and to make efficient use of information. They also free journalists from their desks and landline telephones. Newsplex training director Martha Stone said that a skilled multi-media reporter needed to be well-trained with fundamentally sound skills in writing, interviewing and researching along with a good knowledge of ethical concepts and legal constraints. But in a converged environment they had to appreciate the power of digital tools to help them become better reporters. Journalists did not need to be an expert with all multi-media tools, but they needed to know how each tool worked and how each could be used in the most appropriate circumstances. "We can no longer just sit in our cubicles and be a 'Lone Ranger' writer. We must be aware of how a multi-media operation works, how each of us as a journalist fits into that operation and how valuable we are in that operation. It is indeed a more integrated environment, and it democratizes the newsroom process" (quoted in Roper 2003a: 31).

Stone said new tools also allowed journalists to be more mobile and entrepreneurial. "There is nothing stopping anybody from getting a videophone and going to the city council meeting when they are having another knock-down, drag-out fight, then putting that video online. It shows people how down and dirty a meeting like this can get." People on the streets had captured images of crime scenes with their cell phones. It was relatively easy for journalists to do the same with mobile digital video cameras, she said (Roper 2003a: 30). Ifra writer Mari Pascual said MSNBC.com's Preston Mendenhall represented an example of a successful early adopter of digital tools. Pascual said Mendenhall's trip to Afghanistan in May 2001 showed that converged or multi-media journalism was practicable in terms of available technology (2003a: 34). More details are provided in chapter 4. Chapter 1 relates how John Beeston, CNN's online news director in Hong Kong, sent a woman reporter to Kalimantan in remote Indonesia with a digital video camera, a cell phone and a laptop. The equipment enabled CNN to get more elements of a story compared with a traditional television news crew. "This approach is not always suitable, but again, it gives us more flexibility" (Beeston 2001). In February 2003 the *Manchester Evening News* sent 10 journalists to the first training course in Columbia, South Carolina, after the Newsplex opened. It fired their enthusiasm, even among those most skeptical about convergence. In a video that she produced while there, reporter Sarah Lester said convergence was the future. "We must focus on creating multi-media content offerings that we can distribute on any support. It is possible that the benefits are not immediate . . . but be that as it may, it is vital to move in this direction" (quoted in Pascual 2003a: 34).

Multi-Media Training at the Newsplex

The Newsplex is probably the single best place to learn how to become a multi-media reporter, though dedicated people could train themselves if they had the right equipment plus enough time and motivation. At a typical one-week course at the Newsplex journalists acquire the multi-media mindset needed to operate in a converged world, as well as learn how to package multi-media stories. Training director Martha Stone said reporters enjoyed learning about the gadgets and gizmos. In three full days spent at the Newsplex in September 2004 this author noted a high level of energy as print journalists moved from mono-media to multi-media reportage. The first impression on entering the Newsplex is a sense of light and space. The building is gleaming and white, and egg-crate latticing lets in lots of light but reduces glare. This was not a problem on day one in September 2004 because of all the rain outside, courtesy of hurricane Frances.

The first day began with a video about the Newsplex, in which architect Saf Fahim explained that journalists were never intended "to live in a cave." One of the roles of the Newsplex, he said, was to free journalists. Fahim said he could state, with certainty, "that the environment impacts the way we work, feel, think and react." The act of designing a building puts an architect in charge of people's wellness and being. "Architects know and understand the sheer gravity of this responsibility." Fahim said the newsroom was the "engine room of any media outlet" (Fahim 2004). In a 2001 interview in New York with Australian journalist Sally Jackson, Fahim described some newsroom environments as "not very . . . human." They contained a lot of friction and "negative energy," contrary to what was needed in a creative environment, he said, "which is what a newspaper, believe it or not, is supposed to be." Between 1992 and 1994 Fahim toured a lot of newsrooms across the US after the Associated Press asked him to research ideas to help conceive the design of tomorrow's newsroom. "They were a disaster," he said to summarize what he found. "Absolutely terrible, like a typical office building—offices and cubicles. There's a budget for plants and they give you a plant—an ugly plant. But the fundamental question we asked was: 'Why is there so much separation? Why can't a writer sit down next to a graphic artist and a photographer?' We felt the walls between these tribes had to come down." Walls were everywhere, both physical and metaphorical barriers, he said. Walls had been erected between editors and reporters, between reporters and photographers and all the other editorial "cults," Fahim said. The tallest and most forbidding wall was between the newspaper and other news media (Jackson 2001: 12).

Figure 7. Newsplex founder and Ifra director of publications Kerry Northrup provides feedback to trainees using the news wall. Photo: Stephen Quinn.

The Newsplex has an open and welcoming atmosphere. The walls are white and clean. The area has an air of neatness; equipment fits snugly into discrete panels in the walls, and the bookcases reach to the high ceilings. Northrup said he intended the Newsplex to be "nurturing and caring." His goal was "to get sunlight into the newsroom." The room is filled with technology but it's "not about the technology, but the multimedia mindset" appropriate for working in all media—print, broadcast, online and mobile (Quinn 2004). Training consists of a series of scenarios and exercises relevant to each group of trainees. Northrup said the first day tended to be fairly intense because people needed to get a good idea about the tools available, and the background to the emergence of convergence. As of early September 2004, the Newsplex had attracted more than 2,000 visitors, and just over 220 journalists had been trained. Most were from newspapers.

The first day involved discussion about declining and fragmenting audiences, which were one of the main reasons for the need to shift from media-centric to audience-centric newsrooms. Northrup highlighted the wide range of varied audiences and those audiences' varied use of media across many channels. Trainees identified and described actual and potential audiences for their product. Northrup pointed out that while newspaper and broadcast audiences were declining, the total number of hours of media consumption was increasing about 1.5 percent each year. "Overall media consumption is a growth market," Northrup said, including an increasing number of people who used multiple forms of media at the same time. Chapter 2 discusses this trend.

The group then conducted a SWAT analysis (strengths, weaknesses, assets, and threats) of each of the media platforms. Some of these features are discussed in chapter 4. Northrup noted that when people analyzed the SWAT matrix, they noticed that the strengths of one medium were the weaknesses of another. "So it just makes sense to combine them" in multiple media, he said. The group discussed the process of incremental storytelling that Northrup described as one of the foundation issues in the multiple-media reporting environment. The print mindset said reporters had to have the whole story before they published. With multimedia reporting there was a case for building the story in increments. "As long as you know it's accurate, get it out in the most appropriate medium." Each news story had pieces that people wanted to know about.

Stories consisted of four stages. With the breaking story, news was based on what journalists and their managers knew. "You have to jump on the story at the breaking stage," Northrup said. With the next (developing) stage, reporters had to provide more information, asking themselves what the story meant for the various audiences. With the third stage, which Northrup called the "following" stage, all key facts were known. Now people wanted interactivity and greater detail. "Online, you get few unique visitors but people keep coming back; and this stage may last for days or weeks because the audience is into the story." During this stage journalists provided details of ramifications of the story. In the final stage, audiences knew the story. This stage offered a chance to do analysis, to look at trends, and to conduct chats with people to ask their opinions. Often this stage of the story ran in the newspaper's Sunday magazine. "Audiences are interested here in conclusions," Northrup said, referring to what journalists had learned from the event, "and sometimes audience figures are higher than for the breaking story" (Quinn 2004).

After lunch on the first day Northrup described the new roles discussed in chapter 4: news flow manager, story builder, news resourcer and multi-skilled journalist. Newsplex director Randy Covington introduced a session on how to write for television, before trainees learned how to use Visual Communicator. This was PC-based software with which journalists could construct television packages on their laptops. The software was easy to use and relatively intuitive, and participants were producing broadcast-quality packages within a few hours. The day ended with an introduction to Adobe InDesign, software for creating print documents similar to Quark XPress and Adobe PageMaker. The rest of the course consisted of more scenarios building on the previous scenarios and exercises, as journalists learned more and more tools and techniques for multimedia newsgathering. Along the way Northrup allowed plenty of opportunity to debate issues as they arose. The moblog session where trainees got to practice using camera-enabled cell phones at the local farmers' market proved popular. The results of the journalists' efforts were placed on the web (http://ifraroles.textamerica.com).

Northrup said the moblog was designed specifically to work with a mobile phone. "It's like an online diary, and the chronology is a key part of the process." The moblog was not an ideal medium for covering issues—that was best left to newspapers—but it was fine for describing individual events. Earlier in the chapter Northrup described it as "pointillism applied to publishing," meaning that lots of small events helped audiences understand the complete picture. "The strength is in the whole. Each [moblog image] is a piece of journalism, but you need the total for the whole tone and depth of the story." Northrup described moblogging as "a fragmentary experience" because when people came to a blog they were not looking for *Wall Street Journal* or *New York Times* style journalism, such as an expose. "People come to a moblog for the experience;

Technology and Convergence

Figure 8. Training remains an ongoing need for journalists. These are among the hundreds who have visited the Newsplex. Photo: Stephen Quinn.

they want to see themselves." Trainees began to refer to each moblog news item as a "news-burp." During the Democratic national convention, students back at the Newsplex wrote longer stories to complement the moblog. It was an example of using the strengths of each medium. Trainees practiced linking a longer newspaper or magazine article to a moblog "news-burp," along the lines of how some media cross-fertilize by inserting "for more information" links at the end of news stories. Covington highlighted the need for pre-planning and research before sending people out with camera phones, to avoid the danger of reporters milling around the same subject and duplicating images. "Planning is vital."

As well as experience with moblogs, trainees received plenty of hands-on exposure to other tools—digital video and still cameras, web-based software for managing multi-media content—as well as planning and researching stories for the converged newsroom. Despite the long days—each day started before 9 A.M. and sometimes went after 7 P.M.—the level of energy remained high. The air of intense excitement was tangible. Frowns of concentration appeared on brows as trainees assembled, wrote, edited, and cropped their material into a series of multi-media products. Participants said they felt energized and did not want to stop when offered the chance to go home. Northrup and other Newsplex staff assessed each piece of journalism that trainees produced. Trainees were asked to provide feedback on each tool. Each trainee ended the course by creating a two-minute package using Visual Communicator, in which they synthesized their thoughts on the potential of multi-media journalism (Quinn 2004).

NewsGear and the Newsplex

Each year Kerry Northrup and his team at the Ifra advanced journalist technology project evaluate hundreds of hardware and software products. The most innovative become part of a suite called the NewsGear, which is released at the Nexpo newspaper conference each year and then demonstrated worldwide. NewsGear 2004 represents what Ifra considers the most appropriate tools for cross-media newsgathering. The team aim to keep the cost under $10,000 and in 2004 the total was $9,129. The 2004 edition, which can be found on the web, included:

* Toshiba Portege M200 tablet PC. It has a full-size keyboard and reasonable battery life of about three hours. It weighs just over four pounds, has wi-fi 802, an Ethernet outlet, a 60Gb hard drive, and up to 2Gb of RAM. Advanced handwriting recognition makes it a great tool for the multi-skilled journalist.
* Sony DCR PC330 camera. This camera can take video and still photographs. The latter contain sufficient pixels so that one frame can be reproduced as a newspaper image. It records to a memory stick. The Carl Zeiss lens has a 10x digital zoom.
* Archos AV320 video recorder. The 20Gb hard drive permits recording in several formats: print-quality stills, web video, or broadcast audio. It has a 10-hour battery and a 3.3 megapixel camera attachment, weighs less than one pound, and feeds to Macintosh or PC via a firewire cable.
* Nokia 6600 imaging phone. This second generation camera phone is suitable for real-time newsgathering. The 640x480 camera can take images, audio, and video for breaking news. It is Bluetooth enabled so that reporters have hands free for other duties. It can provide low resolution streaming video over high-speed networks.
* Apple iSight video conferencing system. This item permits desktop video-conferencing so journalists can talk with col-

leagues long-distance, and it switches to permit reporters to record interviews on Visual Communicator Pro. It has firewire for audio and video connections.
* Canon BJC-55 portable printer. This color bubble jet printer allows journalists to print without cables, with a capacity for high-quality printing. An optional replacement for the ink cartridge turns it into a scanner.
* Visioneer Strobe XP100 scanner. This scanner weighs about half a pound and is extremely portable. Reporters can scan documents and images at a news site. It gets power from a laptop, so it does not require its own battery.
* Emergecore IT-100 network appliance. This networking tool offers plug-and-play interface that allows journalists to connect to a variety of networks and servers.
* SmartDisk firewire card. This card allows fast transfer between the Toshiba laptop, the Sony DCR PC330 camera, and the Archos AV320 video recorder.
* DaKine Pod 1 urban backpack. This carrier provides excellent protection for all the tools listed above in an 18x12x8-inch backpack with 1,900-cubic-inch capacity.
* Serious Magic's Visual Communicator Pro. This package includes an advanced version of Visual Communicator plus a Logitech Bluetooth mouse and Zip-Linq retracting cables.

Digital Tools and the Future of the Journalist

CNN's John Beeston, based in Hong Kong, said innovative newsgathering technology would quickly become available to an increasing number of people, and this could nullify the advantage media organizations currently had. "The devices we are using can be purchased by the consumer in most electronics stores in Hong Kong." What, then, could distinguish one news organization from another? Beeston said the issue came down to the quality of the information and the speed with which it was delivered. "All the technology in the world is of little value without content. If we don't maintain and expand the quality of our content, the new technology has the potential to embarrass us."

Beeston said it was easy to become seduced by the possibilities of technology. It was useful only when it served the audience. "So the key is always knowing the audience and catering for them" (Beeston 2001). Al

Tompkins, a broadcast professor at the Poynter Institute, had concerns about the impact technology could have on his industry. "If this [digital technology] gives us significant news in a timely fashion, then it'll be a good thing," he said. "But it's important that the technology doesn't give us tripe and fluff over importance. We don't want expediency to overwhelm journalism" (quoted in Patsuris 2004).

It boils down to this: The brain operating the camera, digital tape recorder, or word processor is much more important than the tools being used. The ability to focus a story is more important than the ability to focus a camera. Before anyone starts shooting or writing they need to know what the story is about. This ability to think critically about a story was crucial to good visual storytelling, noted Regina McCombs, a multi-media reporter for the *Minneapolis Star-Tribune*. "In multi-media storytelling, or any story for that matter, it's so easy to get off track," McCombs said. "It's so easy to find little interesting side things, but when working on a tight deadline you really need to be clear before you know what you're going for. It's a time-saving and clarity issue. It keeps your story clear." McCombs said good reporters should be able to give the gist of their multi-media story in one succinct sentence. The story-in-a-sentence exercise helped journalists focus on the main characters and storyline, and helped them work through the entire storytelling process. "It helps you see your story arc, the structure," she said (quoted in White 2001). McCombs said all the basic rules of journalism applied with multi-media storytelling, but training and education were vital (McCombs 2004). Training prepares journalists to become better storytellers by helping them to focus and plan, and ultimately good training saves time.

7 The Smart Newsroom
Knowledge Management and Convergence

Information is a journalist's raw material. This chapter proceeds from the belief that the better the quality of the information obtained and managed, the better the quality of the journalism produced. It begins by defining data, information and knowledge as they apply to journalism, and then defines knowledge management in the editorial context. The chapter discusses the significance of mindset and attitude in the emergence of convergence, and looks at how newsroom structures and geography influence information flow. The final part of the chapter explores how knowledge management has the potential to provide new information-based business models and sources of revenue. In summary, this chapter will look at:

* Pyramid of data, information, and knowledge
* Defining knowledge management
* The journalistic mindset
* Geography and the flow of information
* Technology and knowledge management
* Establishing a knowledge-based environment
* Promoting knowledge sharing in the newsroom
* Making knowledge management pay

Smart has at least two meanings in this chapter. In the first sense, knowledge management processes can improve the newsgathering process through smarter use of information. In another sense, smart means that media organizations need to recruit, train and retain better-educated staff. The knowledge age needs journalists who can explain an increasingly complex world to their audiences, using a variety of storytelling methods. An individual journalist is also being canny in learning multi-media skills because these skills make journalists more employable, so we could argue that this is a third application of the word smart.

As detailed in earlier chapters, several key changes need to happen for multiple-media newsrooms to emerge. These include a different mindset for managers and journalists, changed newsroom geography, and new attitudes to technology. Managers also need to demonstrate their commitment to change because convergence works best when introduced by managers who demonstrate their involvement. The introduction of knowledge management techniques also enhances the evolution from mono-media to multiple-media newsrooms. This chapter discusses these changes, citing examples of new thinking from innovative businesses and news organizations around the world. It also argues that to be a good journalist involves accepting that learning is a life-long process: Editorial managers must encourage their staff to welcome new forms of knowledge. The move must come from the top, which means that managers must lead by example. They can demonstrate their involvement, for example, by attending training courses themselves. Training and education are the most efficient ways to create proprietary content, and this will be what differentiates great news organizations from mediocre ones. We begin with discussion about various forms of content.

Pyramid of Data, Information, and Knowledge

Every day, journalists gather and process huge amounts of data and information. What is the difference between data, information and the subject of this chapter, knowledge? My explanation involves an analogy. In an earlier life I lived in the Arabian Gulf, in Dubai in the United Arab Emirates, where the locals eat dates at most meals. Data are like dates on a palm tree: They are plentiful, and relatively easy to harvest and measure. They are also unprocessed—data are raw facts—and they only become information when a person uses hands and brain to process them. In the same way, people turn dates into something more valuable such as date honey or date slices. But while dates and other physical products are finite, when we process data we produce more and more information. Chapter 2 demonstrated that this increase in information is becoming a problem for reporters: we daily have to deal with information overload. Indeed, media businesses continue to suggest that information is a glut product because they give it away—witness the daily flood of broadcast news, the mass of free online content, and the number of free daily newspapers and magazines.

Journalists will always be needed to process information. Society needs reporters to sort the wheat from the chaff, to make information manageable. David Shenk, author of *Information Anxiety*, noted that in a world with vastly more information than it could process journalists were important for society's survival. "They help us filter information without spinning it in the direction of one company or another," he wrote. The major challenge for journalists was their willingness to share information, to manage it thoughtfully, and to transform it into knowledge (1997: 166–7). What, then, is knowledge? To continue the date analogy, knowledge is what the scientist creates when she or he reassembles genes to produce a better date palm. It sits atop the pyramid of data, information and knowledge. It is highly refined and valuable, and it is the product of education, training and experience.

Information comes from data, and knowledge comes from information. All involve the human brain but knowledge requires the involvement of people. We find data in records and we find information in messages and reports, but we get knowledge through the interaction of people. In their excellent book *Working Knowledge*, Thomas Davenport and Laurence Prusak showed that knowledge was associated with human beings. "Knowledge exists within people, part and parcel of human complexity and unpredictability." Most people knew intuitively that knowledge was broader and deeper than information, and that it was connected with humans. People talked about a "knowledgeable individual" but it was unlikely that they would refer to a "knowledgeable memo" or a "knowledgeable database" (1998: 5). Knowledge is not neat or simple. When associated with people it is fluid and sometimes unpredictable. It is also intuitive and therefore difficult to capture in words or understand in logical terms.

John Seely Brown and Paul Duguid argued along these lines in *The Social Life of Information*. People were comfortable asking "Where is that information?" but found it odd to ask "Where is that knowledge?" It seemed more reasonable to ask "Who knows that?" One characteristic of knowledge was that it was more difficult to detach than information. People treated information as a self-contained substance: It was something they passed around or picked up, placed in a database, wrote down, counted and compared, or stored and reproduced. Knowledge was more difficult to pick up and move. "It doesn't take as kindly to ideas of shipping, receiving and quantification," wrote Seely Brown and Duguid. Knowledge required more work for assimilation. People digested it rather than simply handled it. The knower had to understand knowledge. This suggested education, work, and commitment on the part of the knower.

People could have conflicting information, but it was rare to have conflicting knowledge. Information could confuse us: We could admit we did not understand information. But that situation was less likely with knowledge. To know something, Seely Brown and Duguid wrote, was truly "to know it" (2000: 119–20).

Defining Knowledge Management

Karl-Erik Sveiby has described knowledge management as the art of creating value by leveraging an organization's intangible assets. He maintained that knowledge could not be managed, preferring the notion of people being "knowledge focused." Sveiby runs a consulting company in Australia, Sveiby Knowledge Management. In an earlier life he owned a series of Swedish financial magazines. "You have to be able to visualize your organization as consisting of nothing but knowledge. This is a different mindset from the industrial era paradigm." Knowledge management involved capturing the two forms of knowledge: explicit knowledge (what can be recorded and reproduced) and tacit knowledge (what resides in people's heads, sometimes called human know-how).

One of the roles of a knowledge manager in a newsroom would be to establish processes for making both forms of knowledge available to staff. In the digital environment, explicit information can easily be copied and reprocessed. The Internet is, after all, the world's largest photo-copying machine. What becomes valuable is what cannot be copied or easily reproduced. Things like integrity, trust, quality, and brand name for a media organization; intellectual firepower, contacts, and experience for an individual journalist. Journalists tend to treasure their tacit knowledge, knowing instinctively that it is more valuable than implicit knowledge. In situations where reporters are reluctant to share tacit knowledge for fear of reducing their own value, another role for the knowledge manager would be to ensure that unique information was securely stored within the walls of the media organization. Contributors need to be rewarded for their willingness to share and assured of the security of their contribution. In the knowledge age, journalists who collaborate make more desirable employees.

In the process of synthesizing information, journalists create knowledge. Over time they generate reputations for integrity and quality journalism. This book has consistently argued that quality content will be the key to a news organization's success in a converged world because it produces, over time, intangibles such as a good reputation and a worthwhile brand. This is what makes outstanding media organizations such as *The New York Times* and the BBC stand out. Some might argue that they have huge budgets and can attract the best staff. I would suggest that excellence is more a state of mind and one of a manager's responsibilities is to foster that mindset.

During the newsgathering process journalists traditionally have changed data into information. One of the key roles of journalism in the forthcoming knowledge age will be to turn information into knowledge—to synthesize it for their audiences. This takes training and skill, both of which require investments of time and money. As Sveiby noted, it took knowledge and energy to convert passive information into something that can be acted on (1996: 386). Seely Brown and Duguid suggested that a shift toward knowledge meant a related shift of focus toward people. As audiences became overwhelmed with more and more information, they would turn to information synthesizers to help them assimilate and understand their world. A true knowledge economy should distinguish itself not only from the industrial economy but also from an information economy. "Attending to knowledge," they said, "returns attention to people, what they know, how they come to know it, and how they differ" (2000: 121). Knowledge management must not merely involve the protection and exploitation of patents. "It must include the cultivation of knowledge workers" (2000: 122).

Davenport and Prusak called the people who synthesize information—librarians, reporters and editors—"knowledge integrators." The title described professionals who extracted knowledge from people who possessed it, put it into structured form, and maintained or refined it over time. Journalists on large newspapers and magazines with specialist rounds or beats fall into this category. Reporters on small publications where everyone is expected to do a bit of everything have less opportunity to be "knowledge integrators" but they can apply the principles in their work environment. Davenport and Prusak lamented that universities did not teach the skills needed to be a "knowledge integrator," though the closest approximation could be found in journalism and library science programs. Consulting firms employed "knowledge integrators"—people with expertise in a particular subject who determined what was valuable and synthesized it. These firms also had "knowledge administrators" who focused on capturing, storing, and maintaining the knowledge that others produced.

Vince Giuliano, CEO of SimVenture, a Boston-based newspaper consultancy, has maintained that newspapers must see themselves as knowledge companies rather than news factories. "While the newsroom is at the center of the newspaper's knowledge capabilities, the newspaper is also a manufacturing company which produces and distributes a new product every day. In terms of costs and economics, a typical newspaper is 30 percent a knowledge-age organization, and 70 percent a traditional manufacturing and distribution organization." The culture and policies of most newspapers reflected the biases of traditional manufacturing companies rather than information-age companies. Giuliano said knowledge-based newsrooms would create long-term possibilities for news organizations, but believed that change ultimately involved transformation of the journalists' perceptions of their roles and profession (2001). In short, a change of mindset is needed.

The Journalistic Mindset

The theme of a changed journalistic mindset occurs most frequently in any discussion of convergence. Several factors are connected to a change in mindset, including a different attitude to time and deadlines, a flexible approach to their journalistic role, a willingness to embrace change, an openness to learning, and a willingness to adopt more collaborative forms of work. As early as December 1999 Kerry Northrup, at the time executive director of Ifra's Center for Advanced News Operations, noted one central theme at a seminar on the type of newsroom needed for a digital age: "the battle over the newsroom's mindset." He concluded that issues of attitude would be "the toughest and most important battle to win in transitioning to a new newsroom." Knowledge management was vital for implementing the kind of newsroom needed for doing multi-platform journalism, he said. This involved an "overhaul" of the newsroom's attitude toward the information it handled and the knowledge it developed (Northrup 1999). John Haile, the former editor of the *Orlando Sentinel* who drove much of the convergence at that paper, noted that managing change was difficult because it was necessary to "change the way your people see themselves and do their jobs."

Northrup said serious implementation of knowledge management produced a major return on investment. "[It leads to] the development of really valuable assets, tangible assets—just like a building or a fleet of trucks—in the form of knowledge bases of information. Newspapers typically are not at [the] forefront of most technologies and true to form they are not at the forefront of knowledge management technologies. But the evolution cycle seems to be a lot faster now than it has been in past decades." Newspapers were recognizing the increasing importance of being an information-based business rather than just a production-based business (Northrup 2000a). Brian Veseling of Ifra's respected industry journal *newspaper techniques*, reporting on the same seminar, said speakers frequently mentioned flexibility as the key factor in successful change in the newsroom. "If there is one word to describe what is being required more and more in newsrooms as journalism moves into the digital age, it is flexibility. In an industry in which flexibility always has been an important element for success, it now seems to be vital for survival." Veseling highlighted the need for changes to mindset and newsroom structures—the latter is what he and others have called "geography"—for knowledge management to flourish (2000: 20).

Geography and the Flow of Information

Intelligent office design reflects the social character of work—the way in which people act as resources for each other, rather than as information providers. Noted Seely Brown and Duguid: "Good office design can produce powerful learning environments. But much of that power comes from incidental learning. For example, people often find what they need to know by virtue of where they sit and who they see rather than by direct communication" (2000: 72). Knowledge seemed to flow with particular ease where the firms involved were geographically close together; hence the various clusters of silicon valleys, alleys and gulches around the world. "Within clusters, there is a shared high-level understanding of the demands and possibilities of a particular sector" (2000: 163–4). Social networks and the physical location of knowledge workers also acted as conduits for innovative thinking. Davenport and Prusak explained that people could not share knowledge if they spoke a different language and would not share if they did not trust each other. The reverse also applied: "The closer people are to the culture of the knowledge being transferred, the easier it is to share and exchange" (1998: 100).

Jan Larsen, knowledge management researcher at Denmark's Center for Journalism Education, said developing trust was "crucial" when managers tried to introduce change among journalists (Larsen 2004). The *Tampa Tribune* has benefited significantly from having all platforms located in one building; proximity has engendered trust. Gil Thelen, the paper's publisher, said that "geography is destiny in this game." Placing all editorial teams in the same building was part of the essence of convergence, along with a centralized news desk as a symbol of convergence, he said. Multi-media coordinator Ken Knight said reporters on all platforms had a greater appreciation of their colleagues after sharing the same building. "That came from working together, with being able to trust each other, and the day-to-day ability to ask questions of colleagues about stories, and know that they will share information even if it is exclusive information. I think that is for the good" (Knight 2004).

Various commentators have described the building Thelen and Knight occupy in Tampa as the "Taj Mahal of newsrooms." It is truly a beautiful building. But the statement also reflects the ugliness of too many other newsrooms around the world: uniform beige computers and carpets, low ceilings, dead or malnourished pot plants, stuffy buildings with dirty windows and towering piles of books and files. The author has visited hundreds of newsrooms in scores of countries. Relative to the ugly and dysfunctional buildings that journalists occupy around the world, the Tampa office is a palace. The offices of the newspaper, tbo.com, and the television station occupy four floors on the west bank of the Hillsborough River, and are filled with the latest digital technologies. "We have nothing analog in the house any more," Thelen said. He referred to a "new energy" in the building. The *Tampa Tribune* had a quicker and better newsroom because of the changes. He attributed this directly to the *Tampa Tribune*'s "editorial muscle" of more than 300 journalists, 120 of whom were reporters. The print publication supplied a sense of security for the television newsroom. As a consequence, the television news produced at partner WFLA-TV was deeper and more contextual. "The daily breaking news collaboration is increasing exponentially and the cultural accommodation is also working well." Thelen noted that an early and important part of his job was changing the mindset in the newsroom and getting people to see that their core journalistic ingredient was news and information. It did not matter in what form they delivered it (Thelen 2000a and 2004a).

Probably the most converged newsroom in the US, in Lawrence in Kansas, operates on the principles of shared knowledge. The newsroom is also a beautiful place in which to work. The combined newsroom for the *Journal-World,* 6News, and World Online formally opened in September 2001. The site was the first federal building in Lawrence. and the stone building served as the town's post office until 1965. Renovations began in 1999. Director of convergence Rob Curley almost gushed about the result. "The place looks incredible. I've been in newsrooms all over the world, and I've never seen one like this." It's a combined newsroom,

Figure 9. The training newsroom of *The Columbia Missourian* at the University of Missouri. Photo: Stephen Quinn.

where a section front from tomorrow's newspaper and sending a breaking news story via a text message (SMS) to cell phone subscribers are discussed in the same meeting. Curley said a sense of history and old-fashioned "newspapering" permeated the building, along with state-of-the-art computers, remote-controlled video cameras and high-speed wireless Internet hubs. The multi-media coordination desk, a symbol of convergence, dominated the ground floor, with desks radiating like spokes from this hub. "The newsroom isn't broken down by medium. It's broken down by beat" (Curley 2004a). Moving into the News Center directly encouraged convergence, Curley said, because it provided an environment that helped generate a collaborative mindset, which in turn became an incubator for ideas. By mid-2004 convergence in the News Center had evolved to the point where it was natural and did not need to be forced. "That's how we came up with the phrase 'organic convergence,'" Curley said (Curley 2004b).

One aspect of the application of knowledge management to journalism involves getting more reporters into the community where they can "brush up against" people. John Hartigan, CEO of News, Ltd., Australia's biggest publisher, has warned that his country was in danger of producing a generation of journalists "who know people only over the telephone—and then only hear the views of spin doctors." Among earlier generations of reporters, he said, "the most venerated journalists found their news among the people . . . you don't find news around the office" (Hartigan 2002: 6). On visits to *The New York Times* in June 2000 and May 2004, I noted the paper's commitment to getting reporters into the community. My host, Stephen Miller, assistant to the technology editor, explained why the vast newsroom appeared so empty at 3 P.M. Reporters were doing what was expected of them: going into the community to talk to people (Miller 2004).

Technology and Knowledge Management

A key change in mindset involves the journalists' attitudes toward technology. They must be willing to embrace the tools that will make their jobs easier, and to adopt those that help in the evolution of knowledge management. But information technology does not create knowledge. It cannot guarantee or even promote knowledge generation in a culture that rejects learning and sharing. Information technology should be seen more as a storage system that permits knowledge exchange. Davenport and Prusak warned that an excessive focus on technology was "the most common pitfall in knowledge management" (1998: 173).

The key remains the mindset—the willingness to embrace technology as a tool for doing better journalism, and to learn how to use it. Ifra's Kerry Northrup elegantly summarized the difficulties inherent in many newsrooms: The late twentieth century had produced newsrooms "full of digital technology—and full of digital problems." Technology has been applied piecemeal to editorial structures, and work flows had not been designed to make the best use of the new tools available to journalists. Too many managers had applied technology to isolated production tasks rather than on the more fundamental issues of managing information flow and newsroom staff. In the rush to introduce digital technology, most newsrooms had not supplied the necessary infrastructure. "Being digital is not just a state of technology. It is a way of thinking, a way of defining the critical processes that make up a news operation, and a way of linking those processes to make the newsroom more dynamic. It means using technology more effectively, not just using more technology" (1996: 4).

Northrup has rightly pointed out that the newsroom of the future must make more astute use of technology. Standardized technology within an organization with a common set of tools and software are vital as a way of ensuring that anyone can read documents anywhere. The most common knowledge management tools are web-based intranets and software such as Lotus Notes. But the presence of the best technology in the world will not necessarily produce a knowledge-creating company. Technology can take information to a user's desk, but it cannot make a person use it. To mangle an adage about horses, you can lead a journalist to information but you cannot make him or her think. However, if managers establish the right climate and encourage their staff to learn how to use the appropriate digital tools, then technology has great potential for expanding and sharing knowledge. Training remains an important aspect of the overall strategy needed to implement a successful knowledge management system. If journalists can learn how to use a word processor or a front-end production system, they can just as readily learn how to run a database or spreadsheet package. The issue is the willingness to learn.

Too much editorial energy has gone into the introduction of technology without relevant training and education. The most important tool editorial managers can give journalists is a climate conducive to knowledge and its transfer. Hank Glamann, a news executive at the *Houston Chronicle,* said the industry's failure to provide adequate knowledge was the "fundamental cause" of technological problems. "We simply do not train our people thoroughly enough to give them mastery over their machines. Is it costly, time-consuming and disruptive to provide extensive training? Yes. Does it pay off in the long run? Absolutely" (Glamann 2000: 48). In 2000, Mogens Schmidt, then director of the European Journalism Center, concluded in a report commissioned by the Dutch Ministry for Education, Culture, and Science that the further training of journalists had become "a permanent need" because the media were "changing and developing at a pace that can only be matched through well-structured continuous training" (quoted in Bierhoff et al. 2000: 5). The report noted that media organizations and journalism schools had generally given priority to skills-based training ahead of knowledge-based learning. That traditional model was "under threat" through being irrelevant. The report called for a new model that offered an academic education (knowledge) as a base, with the practical component (skills) introduced near the end. Regular and integrated training should be offered once people found jobs (quoted in Bierhoff et al. 2000: 12).

Establishing a Knowledge-Based Environment

How do we create an environment conducive to knowledge management and transfer? It must begin with support from senior management. Davenport and Prusak concluded that it was impossible for any company to transform itself into a knowledge-creating organization unless the management team was behind the process (1998: 122). Like money in a mattress, intellectual capital was useless unless it moved around. Managers could establish intranets and other pieces of technology to circulate information. But managers also needed to be able to motivate people to use these technologies. Teams and cross-disciplinary groups appeared to help, provided they had a common cause, project or motivation. Drug companies like Merck give stock options to scientists who invent new drugs. What can editorial managers do to motivate reporters to share knowledge to produce better stories, to bypass the lone wolf mentality? Merck finds ways to reward their scientists if the company cannot produce a marketable drug, by publishing scientific results and adding to the world's general store of knowledge. How can editorial managers work with reporters to re-purpose and recycle information not used in stories?

Tom Stewart of *Fortune* magazine described knowledge management as a process that, like a shrub, needed to be cultivated. Without an established process, good ideas, like seeds, landed on hard ground and were eaten by birds or choked by weeds. Innovation was both a machine and a magic garden, said Stewart. Processes for generating ideas needed to be promoted. "Since innovation is both a machine and a magic garden, keep oiling and fertilizing [it]" (Stewart 2001a). Stewart also argued that committed leadership was vital. Knowledge that evolved from the grassroots level was like grass itself—being low to the ground it was easily mowed. Grass usually spreads by dividing its roots—that is, slowly and only to adjacent territory. Grassroots groups need managers to act as advocates and "scouts," he said (2000a).

Like a lot of companies, media organizations are unsure how to measure or value the intellectual capital their staff generates. An airline company will know how much its aircraft cost, but it is more difficult to put a value on the electronic booking system that puts passengers on those jets. A newspaper's production people will know the value and number of the printing presses, but ask editorial managers how much intellectual capital the paper has, and they will scratch their heads. Managers can put a price on a piece of plant, but can they put a price on the value of a senior reporter or photographer, apart from the salary? What about appropriate development of resources? Newspapers regularly upgrade their production software, or replace their fleet of cars and trucks. Do they encourage staff to further their education? They paint their buildings and promote their mastheads. Do they peg reporters' pay raises to acquiring qualifications relevant to the job? Newspapers recycle newsprint and ink. But what do they do with reporters who retire? How do they make use of that brain that now spends most of its time fishing or making furniture? And how do they ensure that the training that is offered "sticks" and is not forgotten soon after the course? News organizations need to establish appropriate learning environments.

Seely Brown and Duguid concluded that resources for learning were found not in information, but in the practice that allowed people to "make sense of and use that information" and gave access to the practitioners who knew how to use that information (2000: 133). The philosopher Michael Polyani likewise noted that no amount of explicit knowledge provided people with implicit knowledge and experience. "Information on its own is not enough to produce actionable knowledge. Practice is also required, and for this it's best to look to a community of practitioners." As well as mingling with experts and inquisitive people, it's important to accept the value of peer support. Seely Brown and Duguid emphasized that learning was a social process and was demand driven: "People learn in response to need" (2000: 137).

Circulating human knowledge is not a matter of search and retrieval, as some technology-driven views of knowledge management might have us believe. It is easy to use technology to retrieve explicit knowledge and information but it is difficult to detach implicit knowledge from one knower and attach it to another. So the acquisition of tacit knowledge through learning presents knowledge management with its central challenge. The defense of intellectual property, the sowing and harvesting of information, the exploitation of intellectual capital, and the benchmarking of competitors' intellectual assets are all important parts of the knowledge management game. But all of these are subordinate to the matter of learning. It is learning that makes intellectual property, assets and intellectual capital usable. The difficulty of this central challenge, however, has been obscured by technology salespeople, who push technology as a way to provide "solutions" to problems. They define the core issues of information as problems, so that they can offer "solutions," which invariably involve buying more information technology.

Seely Brown and Duguid concluded that this approach did not work. Learning is not simply a matter of acquiring information; it is a social process. They drew the distinction between "learning about" and "learning to be," or the difference between "know that" and "know-how."

Know-how does not come from accumulating information; it comes from practice. "Learning to be requires more than just information. It requires the ability to engage in the practice in question" (2000: 128). Combined with this is the need to develop communities that allow for shared experiences; this can be done through mentoring, brown bag lunches, accidental (informal) knowledge sharing through smart positioning of desks, and social events that encourage people to interact. This brings us to the next section of this chapter: Ways to promote and generate knowledge sharing within a converged or converging newsroom.

Promoting Knowledge Sharing in the Newsroom

This book has argued that quality content or unique or proprietary information generated by knowledgeable staff will be what differentiates the great from the mediocre newsroom in the future. Davenport and Prusak described the three main ways to generate knowledge. The options were to buy or rent it (by hiring specialists or consultants), to grow it (through training, research and development) or to generate it (using methods such as "creative abrasion"). Given the accent on cost cutting at many media companies, money for the first option may be scarce. So this next section looks at the other two possibilities.

Establish Communities of Interest

Knowledge sharing works best within communities of people who trust each other, known as communities of interest. People with something in common are more willing to talk to each other than strangers. *Fortune*'s Thomas Stewart described "communities of practice" as groups that emerged, sometimes spontaneously, around a discipline or problem. "They have no agenda; they are defined by the subject that engages them, not by project, rank, department, or even corporate affiliation. They are where learning and innovation occur—the shop floor of human capital" (Stewart 2001b). Employees who share similar interests or aims form communities of practice. How can editorial managers bring people together to foster these communities? One way to encourage collaboration is to set up environments where people can mingle. The *Maeil Business Daily* in South Korea's capital, Seoul, opened a fitness club. Publisher Dae-Whan Chang said journalists often came to work early to spend time in the club and socialize with colleagues. He found that they

Figure 10. Much tacit knowledge transfer takes place when journalists work in teams, such as at *The Chicago Tribune*. Photo: Courtesy *The Chicago Tribune*.

naturally spent time together discussing work. "Often journalists come to work early to talk about a project they are working on" (Chang 2001). The author has observed gyms and fitness clubs in newspaper buildings throughout southeast Asia but found them less common in the US.

Develop an Atmosphere of Trust

Trust between people is important in the spread of knowledge. "Trust is an essential condition of a functioning knowledge market," wrote Davenport and Prusak (1998: 35). Without trust, knowledge stays in people's heads. MIT researcher Tom Allen found that scientists and engineers exchanged knowledge in direct proportion to their level of personal contact (Allen 1977: 2). Knowledge sharing must be encouraged and rewarded, and it must be well resourced. The serendipitous nature of knowledge must also be accepted—knowledge develops in unexpected ways. Children share knowledge naturally, but over time they are conditioned to become misers. They learn that knowledge is power, so they hoard it. Journalists typically exhibit a deep level of this conditioning.

Editorial managers need to encourage reporters to share information and knowledge. One way is to form reporters into desks or groups (which with time tends to generate trust) and reward the group, not just the individual, for performance. Many Japanese companies have established "talk rooms," where researchers are expected to spend some minimum amount of time each day discussing work over a cup of tea. Dai-Ichi Pharmaceuticals established such rooms and provided green tea and attractive lighting. Company researchers were expected to spend 20 minutes there as part of a normal work day. "The expectation is that the researchers will chat about their current work with whomever they find and that these more or less random conversations will create value for the firm" (Davenport and Prusak 1998: 92). Knowledge exchange works on the principle that people often get information from their neighbors. Davenport and Prusak advocated establishment of knowledge markets, physical and virtual places where people could share knowledge. Providing access to people with tacit knowledge is sometimes more efficient than trying to capture and codify that knowledge. The knowledge market depends on trust, and people tend to trust people they have come to know. What can editorial managers do to give people access to good information and knowledge?

Remember the Importance of Location

Given that ideas tend to be shared in environments where people feel trust and trusted, news organizations need to establish places where people can share ideas. Researchers sometimes employ the term "knowledge leaks" to describe situations where people from different backgrounds gather and sometimes spark new ideas via the meeting of unlike minds. Location remains vital for sharing knowledge: The more widely dispersed the knowledge, the more powerful the force needed to share it. "Magnet locations" bring people together. Organizations need to establish routines and situations where people can get together, for example, by regularly providing a coffee and food cart in an attractive place. Think of the classic idea of the university common room (which still exists at the great UK universities). Provide plenty of comfortable chairs, plants, wholesome food and a stack of books and music, plus other amenities to lure people out of their cubicles. Make it acceptable to "goof off." Establish an "ideas room" in or near the newsroom.

Introduce Change Agents

Change agents are people who advocate for the introduction of new ideas. Davenport and Prusak called them "knowledge evangelists." They help establish a knowledge market place. Innovation theory shows that acceptance of an innovation within an organization is often boosted by internal networks. Theorist Everett Rogers said "subjective evaluations" of an innovation derived from an individual's personal experiences and perceptions that were "conveyed by interpersonal networks"—chats around the water cooler, for example, or in the canteen—will drive the diffusion process (Rogers 1995: 208). One way editorial managers can encourage this process is by appointing a knowledge management leader or "chief knowledge officer." Make someone responsible for the newsroom's knowledge. Give that person an office in the middle of the newsroom and make that office a focus of ideas and information sharing. It could be the location of the "ideas room" mentioned in the previous paragraph. Many American companies, and increasingly those in Europe, have appointed chief knowledge officers to lead the company's knowledge management processes. These companies tend to be consultancies like McKinsey or Ernst and Young. The position goes by a variety of titles: director of intellectual capital at Skandia (Swedish insurance company); director of knowledge transfer at Buckman Laboratories; global director of intellectual assets and intellectual capital management at Dow Chemical; chief information officer at Young and Rubicam (advertising agency); chief learning officer at General Electric. How many news organizations have a similar position?

The roles and responsibilities of these positions are also varied, but Davenport and Prusak maintained that three were especially crucial: building a knowledge culture, creating a knowledge management infrastructure, and making it all pay. "It takes substantial time and effort to create and maintain such an infrastructure, but it's necessary if knowledge management is to become institutionalized" (1998: 116). If firms do not have these three foundations in place and still want to instigate knowledge management process, they should start small and build. In essence, the role of the knowledge manager is to extract and edit knowledge from those who have it, facilitate knowledge structures, and set up and manage knowledge technology infrastructures. They should make sure they measure the value of what they do and, if possible, convert the knowledge they manage into hard figures, such as the cash or time the company has saved because it had a knowledge manager. They should solicit testimonials and references. Chief knowledge officers should be adequately resourced. "If there is one overriding principle to keep in mind regarding knowledge management roles and responsibilities, it is that they should be real jobs requiring dedicated resources . . . [because] few employees will be able to mix corporate knowledge management responsibilities with their existing jobs" (1998: 122). The final section of the chapter looks at ways to make money through knowledge management.

Choose the Right Knowledge Managers

Managers who work with knowledge projects need special skills. They need to be sensitive to people and to understand the "hard and soft" elements of knowledge management. The former refers to people skills and the latter to the technology involved in sharing knowledge. Davenport and Prusak said knowledge managers also needed humility. Tacit knowledge was a sensitive subject, and people who possessed tacit knowledge tended to be sensitive about what they owned. It was easy for those people to resent managers who showed them little respect. "A little humility goes a long way when you're managing a knowledge project" (1998: 113). This suggests the need for a new kind of manager in the cut-and-thrust world of newsrooms.

Understand the Culture

It would be foolish to try to impose methods that work in one culture on another quite different culture. The Japanese idea of "talk rooms" may not work in a culture where talk is considered a waste of time, and people are discouraged from hanging around the water cooler. This is the catch-22 at newspapers where productivity is considered important and people feel uncomfortable if their manager sees them relaxing and talking. In the corporate world of newspapers, people are too busy working to take time off to attend conferences, or reflect or discuss things in a relaxed setting. Davenport and Prusak noted that time was the corporate world's most treasured commodity. "It is the scarcest of all resources, the one impossible to replicate and yet most essential to genuine knowledge creation" (1998: 67). Editorial managers need to redefine their interpretation of productivity to include the time spent in and around coffee shops, bars, restaurants, and talking with contacts. A great deal of tacit knowledge transfer may be taking place. Davenport and Prusak pointed out that tacit knowledge transfer generally required "extensive personal contact" (1998: 95). In the knowledge era, a corporate culture that does not share knowledge generates scarcity. "Knowledge becomes expensive not because it doesn't exist but because it is hard to get," Davenport and Prusak noted.

People may lack the knowledge to do a better job because of a hoarding culture. The answer in this situation is to change the culture. One way is to reward people for sharing. Davenport and Prusak reported "direct evidence" that promotion, rewards or recognition for generating knowledge sharing were worth the effort and time spent. Rewards for journalists who embrace multi-media reporting are equally worthwhile. But organizations need to be willing to pay: "To establish a consistent culture of knowledge sharing, you need to use valuable currency: substantial monetary rewards, salary increases, promotions, and so forth" (1998: 47–8). McKinsey spends 10 percent of its revenues on knowledge management. Ernst and Young allocates 6 percent. By contrast, media organizations tend to be misers in this area.

Establish a Mentoring System

Mentoring is one way to share tacit knowledge between an expert and a learner. The best way to learn writing or editing is at the shoulder of an experienced editor. One of the guiding principles should be the idea of showing as well as telling. Thus trainee editors should be able to watch the screen of a senior editor working on a story. Mentoring requires a level of altruism on the part of the mentor. Gail Sheehy writes that people go through stages in life, and enter a stage when they find it important to pass on what they know, usually in their fifties. Can this happen at news organizations that retrench journalists soon after they reach 40? Media organizations need to guard against the loss of corporate memory. This is itself part of knowledge management. Knowledge also takes time to evolve, the way a good wine matures in the cellar. It comes from "experience," which has the same origin as the word "expert." Both words come from the Latin root to "put to the test." Experts are people with deep knowledge who have been tested and trained by experience. Knowledge born of experience can make connections between what is happening now and what has happened in the past. "When firms hire experts, they're buying experience-based insights," Davenport and Prusak noted (1998: 8). Knowledge also needs to keep evolving to be effective. Knowledge that has stopped evolving eventually becomes dogma. Knowledge contains judgment—the ability to see clearly—while data and information do not. Training and education are vital to keep people's knowledge fresh. If we are not expanding, we are contracting.

Establish Knowledge Maps

Knowledge maps show people the location of knowledge that an organization considers important, and what knowledge is available. They need to be prominent and highly visual. Information needed to establish a map is often available within a company, but in an undocumented and fragmented format. Different people have different forms of information and knowledge in their heads. The task of an organization's chief knowledge officer is to combine these mini-maps into one knowledge map. Fact checkers, researchers, and librarians at media organizations are good at this pooling of resources, and are the ideal people to approach for advice.

The "news resourcer" discussed in chapter 4 would be another resource. Maps work best when updated as often as possible. Establishing an electronic Rolodex for the newsroom should be one individual's responsibility, and widely publicized to encourage people to contribute. The person responsible for this electronic Rolodex should have sufficient credibility within the newsroom to command the loyalty and respect of even the most hardened lone wolf reporter. That individual must have the power to demand information, if necessary, and should designate a deputy to take over when the leader is absent, to ensure continuity. The issue of getting lone wolf reporters to share the contents of their treasured and dog-eared contact books must be dealt with, perhaps through a process of rewards and flattery. It is their unique tacit knowledge, after all, that they are being asked to surrender. Davenport and Prusak pointed out that knowledge maps made knowledge easier to find and also promoted the idea that corporate knowledge belonged to the organization as a whole and not individuals. Digital maps have the advantage that they can be updated and distributed more frequently and easily than paper versions.

Hire the Best People and Pay Them Well

The most important resource for developing knowledge is an adaptive work force: people who can acquire new knowledge and skills easily. The best predictor of mental nimbleness is proven experience in taking on new tasks. Davenport and Prusak said firms should seek out employees who had already mastered a variety of roles and skills. They also recommended "learning sabbaticals" to allow people to master new work-related disciplines (1998: 65). After they've been hired, employees should also be encouraged to change jobs often, to build and manage their skill portfolios. But essentially the key is hiring the most appropriate people. How many editorial organizations use anything beyond an interview for selecting people? The vice president for electronic media for the Newspaper Association of America, Randy Bennett, said the challenge for newspaper companies was "to find, recruit and retain technology-savvy personnel" (Bennett 2000: 4). Are your newspapers hiring the most intellectually curious? Are salaries commensurate with those expectations? Are reporters encouraged to study to develop their intellectual capital?

In South Korea the Maeil Business News Group, which publishes the *Maeil Business Daily*, provides financial aid for reporters who enroll in a doctoral program abroad. Reporters receive special study leave, and "there is no restriction to reporters' leave for further study," publisher Dae-Whan Chang said. The company also runs a Knowledge Management Academy for business executives in conjunction with KAIST, the Korea Advanced Institute of Science and Technology.

KAIST's Graduate School of Management runs one of the best business schools in Korea (Chang 2001).

Use the Orientation Program Wisely

The orientation of newcomers is an excellent opportunity to start the knowledge management process. Foster an appropriate atmosphere early. Basic training works in the army; newspapers should learn from it. Get established staff to list their strengths and weaknesses in writing. Set goals. The act of writing and recording compels people to think about what they know and what they need to know about an intended story or project.

Balance Openness and Privacy

It is important to ensure that people have a place to work in private yet also feel they are part of an open-plan environment that allows the sharing of information. Executives at some organizations have glass-walled offices; at some innovative offices the walls can be written on and pens are available everywhere, so that if someone has an idea they can write it on the wall. There it is open to comment from others (a less smutty version of the conversations that take place in public toilets around the world). Saf Fahim, the architect who designed the Newsplex, emphasized that it was important to balance privacy with an open plan environment where people could share. "People have to create a personal space that they can call their own. This is very important and elemental. . . . In my opinion these spaces should have a meditative quality to them. That typically stimulates the creative energy." Fahim said that when he created places where people would "truly interact" he had to think about the "social quality" of the newsroom. "For example, we gave considerations to issues like how to break boundaries and how to eliminate territorial instincts. This was most difficult because these spaces not only had to be functional for multiple individuals and tasks, but also the space had to instigate collaboration among people" (Fahim 2004).

Editorial managers similarly need to provide places where people can mix in large, open areas but also establish offices where people can work in private to concentrate on deadlines. Cubicles may not be the most efficient way to organize a newsroom. Allocating offices to senior editorial managers that keep them separate from their staff appears a foolish allocation of space. Section editors at *The New York Times* each have two offices: A separate room where they conduct private business, and a desk in their area of the newsroom where they interact with staff. Robert Whitehead, editor of *The Sydney Morning Herald*, has no walls on his office, which is located in the main newsroom. Jan Larsen of Denmark's

Center for Journalism Education said that the balance between privacy and openness when designing knowledge management systems in newsrooms was a vital issue (Larsen 2004).

Bring in Visiting Speakers and Experts

Ikujiro Nonaka and Hirotaka Takeuchi wrote one of the seminal books on knowledge management, *The Knowledge-Creating Company*. They noted that bringing together people with different knowledge and experience ("creative abrasion") was one of the necessary conditions for problem solving. Differences between people from a variety of backgrounds stopped the group from adopting familiar patterns or settling into grooves during the process. Davenport and Prusak noted that a "significant commitment of time and effort" was needed to give the group enough shared knowledge and language—vital ingredients for success—to be able to work together (1998: 61–2). "Managers should regard communities of practice as company assets and look for ways to preserve them," they said (1998: 39). One way to create abrasion is to bring in outside thinkers, to invite researchers and members of think tanks as visitors to the newsroom. Let those people spend time among the reporters or attend the daily news meetings, and give visitors and journalists a chance to mix and talk. Many academics would be glad to have a chance to mingle with journalists in the newsroom, especially if the academics were asked about their areas of expertise.

Use Databases Wisely

Databases are excellent knowledge management tools. Xerox's Eureka database is a celebrated online collection of fix-it tips prepared by the company's copier repairmen (itself a community of interest). About 15,000 Xerox technical staff employed it during a quarter-million repair calls a year. This shared knowledge in Eureka is said to save Xerox about $11 million a year. Eureka is an example of vernacular knowledge sharing: the harvesting, organizing, and distribution of insights that come from an organization's grassroots. Toyota's Production System and GE's WorkOut are based on the same premise: the collective knowledge of ordinary production workers is valuable. How can editorial managers assemble the collective knowledge of their reporters and editors?

Making Knowledge Management Pay

The person responsible for knowledge management within an editorial company needs to find ways to make knowledge management pay, either by showing how knowledge management saves money or time for the company, or by generating revenue. The key from the outset is the quality and integrity of the information and knowledge generated. So relevant legal issues such as intellectual property—knowing who owns what knowledge—should be sorted out from the start. News organizations also need to establish codes of ethics that describe appropriate ethical behaviors. A reputation for quality is difficult to establish, but easy to lose. In an information-glutted world, people will gravitate toward brand names associated with quality, so issues of ethical behavior are paramount. Knowledge management gives media organizations a chance to position their brand at the quality end of the market. Jan Larsen of Denmark's Center for Journalism Education said knowledge management might not make a media organization rich but it could help some of them grow more effectively and ultimately could influence their capacity to survive (Larsen 2004).

People will only pay for materials they consider useful, so one of the first jobs of the chief knowledge officer should be to identify the relevant audiences for knowledge and find out what they want by asking them. The chief knowledge officer also needs to think strategically. What are the relevant business models? What are the organization's core competencies? What are the most appropriate technologies? How do we archive information so that it is accessible in a variety of ways to a variety of people? Is it XML (extensible markup language) compatible? Who will do the tagging—the application of key words—so information can be cross-referenced? When and where will tagging take place? Digital equipment needs to be continually updated. What specifications are in place to ensure ease of update? Is the company's software accessible and easy to update? This next section offers specific ways to generate revenue, though these should be seen more as case studies from which to extrapolate ideas.

Recycle Timeless Material Intelligently

All media organizations have stories that could, with some thought, be repurposed into useful new products. Examples include research documents to help people make better decisions. For example, many media organizations review consumer items such as cars, wines, or DVD players. Most people seek advice when buying major consumer items. Why not compile and sell documents that include recommendations culled from reviews? Obviously, these documents would increase in value depending on the cost of the items, so a guide to buying a house or a car would be worth more than a guide to the best cell phone or microwave oven.

Think Thin, and Then Think Creatively

What parts of the editorial knowledge base can be made available using technology that allows us to send a thin slice of information to an individual? The mother of a friend in regional New Zealand buys her daily paper solely to read the obituaries. My response was to suggest an email-delivered "daily death." She would be sent only the obituaries, either daily or as a composite each week or month. That's thinking thin. Now we need to think creatively. Extrapolating from the "daily death" scenario, who else would buy such a publication? Perhaps the people in a city like New York who were desperate to rent an apartment and who needed advance warning on available accommodation? Or funeral directors keen to drum up business? Or security companies anxious to sell their services, because they know that burglars often read obituaries to know which apartments will be empty and therefore easy to break into? Or police anxious to catch those criminals? The issue here is creativity. A course in Edward de Bono's lateral thinking would not go astray in most newsrooms, where thinking patterns tend to be linear.

Think Backwards and Forwards

Researchers are also obvious consumers of stored knowledge. They need historical information, in textual, image, or moving-footage form. The issue here is how the information is stored so that researchers can access it, and pricing the information so that researchers can afford it. Academics and school teachers and students cannot pay high prices. These are the kinds of people who need media information for creating timelines, today-in-history lists, complete with photographs, text and broadcast clips. Timeless material in the features sections of newspapers and magazines can be recycled many years later. Readers may be interested in finding what fashions, recipes, or cars people used several decades earlier. Sports fans may want to see video of the best football goals for the previous year, or the best basketball action photographs. To make the information we use now more available in the future, we need to be thinking of ways to store it adequately. The issue here is making all forms of archived information manageable.

Converged media organizations need to think and act more intelligently. They need to employ more creative people who can operate in a changing environment. In short, media organizations need to be smarter and more flexible. Knowledge management combined with convergence plus the ability to hire smart people will ensure that media groups will survive and flourish in this new century. What will the future look like? In the next chapter we take out our crystal balls and look into that future.

8 Convergence and the Future of Journalism

The lessons of history show that the availability of improved news-gathering technology changes journalists' information gathering and reporting methods. Reporters are forced to learn new skills. At the same time, their managers have to learn how to deal with change as well as appreciate the significance of the new tools. The media industry also adapts to cope with changes in society as part of its role of reflecting that society. Convergence is one of the biggest changes that journalism has had to confront. This chapter considers the likely changes to the profession, and includes discussion on:

* Historical notes on technology
* Ages of technology in relation to journalism
* Silly predictions by academic "experts"
* Digital technology, especially broadband, improves
* Adapting to changing consumer habits
* Uncertain regulatory frameworks
* Impact of innovative technologies
* Impact on journalists
* Content still the key

To see the future more clearly, sometimes it helps to look backwards. We will begin with the aroma of coffee in London more than 300 years ago. Journalism began as word-of-mouth delivery of news. Coffeehouses in the seventeenth century evolved into a lively forum for the exchange of news and gossip. These institutions flourished in England and Europe before and after the development of the newspaper. Some historians say newspapers evolved out of these coffeehouses. Media historian Mitchell Stephens has noted that much of the history of journalism can be understood "as a long struggle by written and printed forms of news to compete with that first news medium—word of mouth" (1989: 165). By the start of the eighteenth century London had hundreds of coffeehouses and each house developed its own character. "Lloyd's attracted ships' officers, traders, merchants and bankers, and specialized in shipping news. . . . Will's coffee house was known for its collection of 'wits,' including John Dryden and Joseph Addison" [the latter founded *The Spectator*]. Stephens concluded that London had so many coffeehouses that, like modern magazines, "they were forced to seek out narrow audiences" (Stephens 1989: 42–3). Lloyd's emerged as a company that specialized in shipping information, publishing a daily newspaper called *Lloyd's List* (founded in 1734).

It could be argued the "cyber-cafes" that appeared in the late twentieth century provided a link with earlier times, in the sense that people still congregated to drink coffee and talk, except they also surfed the web. The main difference was that they could gossip with someone across the world rather than across the table. Digital technology perhaps shows its roots in such coffee-based puns as "java" and "mocha" used to describe software. British media academic Brian McNair said new technologies such as the Internet could create a new version of the "coffee-house culture" of the eighteenth century, allowing readers "to escape from the clutches of media barons such as Murdoch, Berlusconi, Springer or Black" (1998: 142). Journalist Anthony Sampson wrote that the "best elements" of the British media had always arisen from natural discussion, curiosity and questions: "If mass communication has become too distorted and corrupted, it may be thought that the Internet and email will provide the new technologies to rescue [audiences] from the old ones, to build up more reliable systems of information across the world" (1996: 51). Coffeeshops and online cultures recognize the human need and desire for community. The ability of the news media to create and maintain a sense of community will directly influence their ability to thrive in the twenty-first century.

Historical Notes on Technology

This book has considered how the availability of digital technology is changing journalism. Looking at some of the earlier technologies that revolutionized journalism may offer a sense of what may happen in the future; indeed, to learn the lessons of history. Winston Churchill observed that the person who ignored history was condemned to repeat the mistakes of history. What can the early history of newsgathering technologies teach us as we move into the twenty-first century? Journalist Tom Standage who writes for *The Economist* called the telegraph the Internet

Convergence and the Future of Journalism

of the Victorian era (1998: 1). The development of the telegraph in Britain and the US followed the growth of each nation's railway system. Newspapers and journalism in turn developed because of the increased amount of information the telegraph made available, and the rise in public interest in domestic and international events.

Railways also helped distribute newspapers. Media historian Jeffrey Kieve noted that the "stimulus" for the telegraph in each country was the expansion of the railway (1973: 13). A boom in railway development was a feature of 1830s Britain, and a similar surge occurred about a decade later in the US. By the 1840s, British investors had invested a sum equivalent to 10 times the then value of the country's imports in rail networks. In the US, the telegraph similarly followed the rail networks. Samuel Morse's eponymous code was first used with the telegraph in the US in 1844, and it remained the basic form of telegraphic communication for more than a century (Pownall 1973: 5–6). The news industry of the mid-nineteenth century thus grew hand-in-hand with the key technologies of that era: steam railways and electricity.

The news industries of the twenty-first century are also products of the dominant digital technologies of this era: the web, email, broadband, and third-generation cell phones. Between the mid-1840s and the American Civil War (1861–65), the telegraph transformed American journalism into a news-hungry industry. News became something that was topical rather than what was reprinted from overseas newspapers when they arrived, usually months later. Broadband and a host of emerging technologies such as third-generation cell phones will similarly transform journalism over the next decade.

The telegraph achieved significant fame in England in 1845 because of a major news story. In January 1845 John Tawell murdered his mistress in the town of Slough, about 15 miles west of London. Slough was one of the stations on the Great Western Railway recently equipped with a telegraph. Tawell fled by train to the anonymity of London, dressed as a Quaker, and police arrested him when he arrived at London's Paddington station. Jeffrey Kieve said that the transmission of Tawell's description by telegraph to Paddington "was largely responsible for his rapid arrest." Publicity around the arrest heightened public awareness of the new device and the telegraph became famous as "the cords that hung John Tawell" (1973: 39). Newspapers in Australia highlighted a similar incident to mark the opening of the telegraph that linked Portland in western Victoria to Melbourne in 1858. They published a story from a Tasmanian newspaper about the capture of a prisoner who had escaped from Hobart only to be caught in Launceston, 90 miles to the north of the island.

Historian K. L. Livingston said police arrested him because of the description provided over the telegraph (1996: 53). In the early twenty-first century, modern communications technologies such as satellite phones, laptops, videophones and digital cameras have boosted journalists' reach to a degree that their nineteenth-century counterparts could never dream of. Will major news stories such as the Iraq war, the Abu Ghraib prison horrors and the tsunami of Boxing Day 2004 be seen in future years as examples of the power of digital tools for reporting?

To learn how journalism may evolve in the new millennium, one should look for modern equivalents of those major nineteenth-century infrastructure developments: the railway, electricity, steam, and the telegraph. Modern equivalents are likely to be some combination of third-generation cell phones, satellites, digital cameras, and broadband. The Abu Ghraib prison torture story offers one example. Senior BBC journalist Nik Gowing said America's reputation in the Arab world had been "catastrophically damaged" by the images coming out of the prison. They had been smuggled out of the prison on a memory stick the size of a thumbnail. That technology could "challenge the power of a government," he said. "Think of something like the image of an Iraqi prisoner in the American-controlled Abu Ghraib prison in Iraq being led around on a leash. Think of that image on the front page of a newspaper under a headline that screams 'Treated like a dog.' When you think of the power of that memory stick, that image, and the impact it has had on the image of the United States, particularly in the Arab world, it spells disaster." Gowing said the spread of digital photography combined with high-speed and low-cost communications was fast transforming the nature of the media and its relationship to government (quoted in *Business Times* 2004).

Ages of Technology in Relation to Journalism

It is instructive to look at the various newsgathering tools that journalists have used over the past two centuries, related to what we could call ages of technology. It is important to note that a significant degree of overlap occurs in these ages, especially as the time gets closer to the present. Historians generally agree that the period from the 1840s to the 1920s was "the age of the telegraph." One significant consequence was the development of many newspapers, boosted by improvements in press technology and steam engines, the spread of railways, and the start of news agencies such as Associated Press in 1848 and Reuters in 1851. When telegraph lines began to spread significantly, it became clear that old newsgathering methods, such as one correspondent sending a report by the best available means to one newspaper, could not survive. The newspaper that controlled a telegraph line had a decided edge over its competitors. The spread of the telegraph also influenced employment demographics. More women entered the work force as telegraph operators—and later typists when the typewriter became a common business tool—because their smaller hands and fingers made them more suited to the intricate work.

The 1920s to 1950s could be considered the age of the telephone because long-distance calls became available and relatively common from about the late 1920s. At the time, journalists used such terms as "telephone-assisted reporting" to distinguish news reports not gathered through face-to-face contact. During the 1980s journalists began to use technologies such as the Internet, databases, and spreadsheets, and a form of journalism known as "computer-assisted reporting" evolved. In a similar fashion to telephone reporting, over time the phrase "computer-assisted reporting" will disappear from the modern lexicon. The telephone led to the development of radio journalism and the early broadcast industry. American journalism academic Randy Reddick believes the telephone had a profound impact on American journalism in the early twentieth century because it enabled journalists to collect information more quickly and conveniently. Reporters at the news scene would telephone reports to rewrite desks, shortening the time between an event and the printed article about it. A new job classification, known as the rewrite person, subsequently emerged in the US, and as telephone technology improved, so its impact on journalism continued.

In addition to phoning information to newspapers, reporters could call sources. Journalists could locate people across the country or even the world. Telephone interviews became a standard reporting method. Reddick said the Internet of the mid-1990s had evolved to about the same level of development as the telephone systems of a century earlier. "There is no universal access. It is not easy to operate. And only a select number of reporters currently use it" (Reddick and King 1995: 71). A decade later, that situation had changed to the point where the bulk of American reporters had a personal email account and searched the web extensively for information. We are also seeing a return to an old idea with the introduction of the continuous or advanced news desk at major American newspapers, which revisits the notion of the rewrite desk (chapter 5 has more details). And as cell phones and personal data assistants

(PDAs) develop we could see the emergence of new forms of newsgathering, as well as new ways to distribute news and information.

By 1915 all European capitals were linked by telephone. A generation later, World War II produced significant developments in computer technology because artillery needed accurate ballistics data. Previously artillery charts had been calculated by people, who were actually known as "computers." Developments in technology during World War II presented a classic example of newsgathering technology evolving more quickly as a by-product of military innovation. Examples include the communications satellite, which gave journalism the video-phone and the satellite phone. We could argue that military technologies also boosted the computing developments that led to the evolution of desktop computers, laptops, and laptop-based editing software, along with a host of communication tools like the cell phone. The period from the middle of the 1950s to the 1980s has been called the golden age of television, and from about the 1950s we entered the age of computers. Digital technology, the child of analog, is now revolutionizing journalism. We will now discuss the future that this digital technology is producing.

Silly Predictions by Academic "Experts"

Before we do, a word of warning about academic experts. It was always difficult to make predictions, the famous baseball player Yogi Berra once said, especially predictions about the future. Some famous examples related to the media make this point:

* the radio industry would make its money from selling radio receivers rather than through advertising
* once it built up huge audiences, radio would kill off newspapers (the newspaper business subsequently boomed)
* television would lead to the demise of cinema and/or radio and/or newspapers
* pay TV would never make money, given the amount of free-to-air television available by the 1970s

My crystal ball is cloudy, but we can use the hindsight of history to make some predictions, based on some recent developments in society and technology. These developments include vast improvements in digital technology (especially broadband), further changes in the way that consumers want their news, and continued uncertainty about the regulatory environment. Let's look at each of those.

Digital Technology, Especially Broadband, Improves

The Pew Internet and American Life Project in 2004 reported that two in five people in the US accessed the Internet via high-speed links at home. Pew estimated the adult residential broadband population was about 48 million, or a quarter of all adults. Among college-educated adults aged 35 and younger, penetration had reached 52 percent. Internet availability at most businesses has become almost universal. Broadband is available in several forms. Digital subscriber line (DSL) from telephone companies and cable modems provided by cable television companies represent the largest chunk of the market. The rest comprise a range of technologies including satellites, fiber-optic, and wireless. As of March 2004 DSL made up 42 percent of the home broadband market and cable modems maintained a market share of 54 percent. DSL connections cost about $30 a month and offered average speeds of between 500 kilobits and 1 megabit per second. This was fast enough to surf the web, swap photos, and download music and brief video clips. Cable modem broadband connections cost between $40 and $45 a month and had speeds of about 3 megabits a second. This was acceptable for watching video and fast Internet surfing but still not fast enough for things like movies on demand (which would need speeds of about 15 megabits per second). Internet connections at speeds of 15 to 20 megabits a second will not be available to most American consumers until about 2007.

Pew predicted that 60 million Americans, or about a quarter of all households, would have some form of broadband connection at home by 2008. The Nielsen research company was more optimistic. Based on current growth patterns, it said the number of households with broadband would rise from 40 percent in 2004 to 62 percent by 2008 (NielsenNetRatings 2004). Regardless of which data we accept, the point is that broadband users have a more intense relationship with the Internet and other media compared with dial-up users: Seven in 10 people with home broadband go online daily versus about half of dial-up users. Two in five people with home broadband get their news online, compared with one in five dial-up news consumers. As New York Times Company chairman Arthur Sulzberger Jr. noted in the opening of the first chapter, broadband changes media consumption. Broadband will be a driver of convergence because it is an always-on delivery mechanism. It makes newspapers attractive to potential convergence partners because newspapers have larger numbers of reporters to provide strong local newsgathering capacity. They also have good relationships with advertisers built up over many years, and a thorough understanding of their market and environment. Broadband also allows for delivery of news, information, and entertainment via multiple devices. Disney's ESPN network has a product called ESPN motion that makes high-quality video clips available online. ABC News, owned by Disney, is using the same technology for online distribution of news (Rosenbush 2004).

Media analysts have noted major changes in how people use media because of the availability of broadband. South Korea is the most wired country in the world in terms of broadband connections, with 71 to 73 percent of households connected as of early 2004. The uptake of technology has been the direct result of government policy. In 1995, perhaps 1 percent of South Korean residents used the Internet. The decision to focus on broadband essentially started after the collapse of Asian financial markets in 1997 devastated South Korea's economy. Policy makers chose technology as a key for restoring the country's economic health. By the end of the twentieth century, 30 percent of households had broadband access through digital subscriber lines or cable modems. The government spent $24 billion building a national high-speed backbone network linking government facilities and public institutions. Most Koreans paid $27 a month for a connection speed of between 3 and 8 megabits a second. Koreans in major cities tend to live in large apartment buildings. These buildings make it is easier and cheaper for telephone companies to deploy high-speed broadband, compared with the US. By late 2005, more than 80 percent of Korean households were scheduled to have access to connections of 20 megabits a second, sufficient for high-definition television.

In March 2003, *New York Times* reporter Howard French wrote that more and more South Koreans were getting their information and political analysis from "spunky" news services on the Internet instead of from the country's "overwhelmingly conservative newspapers." The best-known service is OhMyNews.com, which launched in February 2000. The founder was journalist Oh Yeon Ho, who turned 40 in 2004. OhMyNews.com is based on a popular phrase in South Korea and is not named after Oh. "I launched OhMyNews.com on 22 February 2000 at 2.22 pm. That was my farewell to the journalism of the twentieth century," he said. During the 2003 elections the free service was recording 20 million page views a day. The numbers later dropped to about 14 to

15 million visits a day, still remarkable in a country of about 48.5 million people. OhMyNews.com is unique in having somewhere between 20,000 and 30,000 "citizen journalists" who contribute stories on everything from local events to national politics (Han 2004). These "citizen journalists" are a fascinating subject, and perhaps a prelude to new styles of journalism and newsgathering.

As of mid-2004 OhmyNews.com employed 25 trained reporters who covered the major news stories of the day. Another 10 editors reviewed and posted as many as 200 articles a day written by the "citizen journalists." The country's geography—South Korea is about the size of Indiana—made news coverage manageable because it was relatively easy for one of the site's professional journalists to visit or contact a citizen journalist. No one was more than a few hours away by car or train. Contributors also accepted a code of ethics and agreed to avoid discussing pornography or racism. "They are writing articles to change the world, not to earn money," said Oh. *Newsweek* reporter Christopher Schroeder said OhMyNews.com had become one of the most influential news and information sites in Korea. Oh said OhMyNews.com was a "special product of Korea" because Koreans had little access to open and free dialogue and a large portion had grown dissatisfied with the mainstream conservative media. In the early years of the new century, the combined circulation of the country's main newspapers dropped by a third. Oh said South Korea's high broadband access had allowed people to become more engaged. "[But] technology itself cannot change society. Korean citizens were ready to participate. Only prepared people, who can use the merits of technology, can make a difference" (quoted in Schroeder 2004). Are American journalism programs preparing students for this new world? And are American journalists and media organizations preparing themselves for the changes that will come with broadband technology?

Adapting to Changing Consumer Habits

Audiences have changed markedly over the past generation as media organizations have fragmented and adopted new forms. When audiences change, journalism and journalists have to change to accommodate them. In the early twenty-first century, audiences want news when it suits them, rather than when the media have traditionally supplied it. Howard Tyner, a former editor of *The Chicago Tribune,* retired at the end of 2003 as vice president/editorial for Tribune Publishing. He has long maintained that

Figure 11: An aerial view of the multi-media desk at the Tampa News Center, taken early in the day. Photo: Stephen Quinn.

the business of journalism was about "eyeballs"—getting as many people as possible to look at media products. "Our business was and still is to gather the news, process the news and then distribute the news to consumers or customers or whatever you want to call them. My feeling is that journalists need to be nudged into the notion that their job is news not newspapers. The newspaper is and will be for a long time the engine to gather and edit news. But it won't be enough to just deliver that information to newspaper readers. We need for our news and information to go to the eyeballs of web consumers and TV viewers and cable customers" (Tyner 2004).

Gil Thelen, publisher of the *Tampa Tribune*, said people's information-seeking behaviors were changing and media organizations needed to adapt to respond to those behaviors. In October 2003 the BIGresearch company of Columbus, Ohio, reported significant simultaneous use of multiple media, with three in four people watching television and reading the newspaper at the same time. Two in three people regularly or occasionally watched TV while going online (API 2003). Ruth de Aquino, a newspaper manager in Brazil, pointed out in the first chapter that the public's consumption of news in the new century has changed dramatically compared with the early 1990s. News and information were available on a huge array of devices and many other avenues had yet to be discovered, she said. The concept of news was changing and becoming more personalized, more service-oriented and less institutional (de Aquino 2002: 3).

If journalists intend to reach the largest number of possible audiences, then convergence makes sense because it allows for the widest possible reach. Ifra's Northrup said the fundamental shift we could expect in the news and information marketplace was from mass audience to aggregated audiences. "Overall news and information consumption is increasing an average of 1.5 percent a year, plus more people are tending to use multiple media at the same time. Yet market share for most individual media is flat or declining. To tap the growth market, therefore, media companies must expand convergently" (Northrup 2004d).

Uncertain Regulatory Frameworks

Chapter 2 offered Singapore as an example of the influence of regulatory factors in determining media structures. To recap: until 2001, two companies maintained an effective duopoly. In 2001 Singapore Press Holdings (SPH) owned 11 daily newspapers—six in English, three in Chinese, and one each in Malay and Tamil—and controlled all print news. At the same time the Media Corporation of Singapore (MCS) operated five television and 10 radio channels. In 2001 the laws controlling media ownership changed and in November that year MCS started a tabloid daily, *Today*. SPH subsequently launched two television channels and trained many of its print journalists to supply the content for those channels' news and current affairs programs. ACNeilsen reported that the media consumption habits of the country's population were "constantly being reshaped by the changing media landscape in Singapore" (ACNeilsen 2003). One of the uncertainties that media organizations around the world will have to contend with is change in laws that affect ownership or media structures. As of the time of writing, the situation in the US was uncertain.

Impact of Innovative Technologies

We should never underestimate the power of innovative technologies and tools. Technologies such as electronic paper (e-paper) and distribution via the web or a similar digital form loom on the horizon, as do more innovative forms of communication such as multi-media messaging. Electronic paper and digital distribution are attractive in terms of saving money—newsprint and distribution remain two of the biggest costs for a newspaper. It remains unclear how consumers will react to e-paper and the prospect of reading news in digital form. What is clear is that this is not convergence because it involves the same product (newspaper editorial and advertising) reaching the consumer via a screen instead of paper. What is also clear is that the print and broadcast industries need to find ways to pay for convergence and other new forms of media.

Andrew Nachison, director of the Media Center at the American Press Institute, believed the future of news was mobile. The challenge for news organizations was how to deliver the product to an increasingly fragmented market. Personal digital assistants and sophisticated cell phones were already outstripping newspapers, television and even the Internet as a primary source of information in parts of Asia and Europe. Nachison predicted that wireless cell phones would become powerful, multi-media devices that would replace traditional personal computers. Niklas Jonason of Citygate in Sweden urged publishers to think about cell phones as ways to reach younger demographics, especially as the technology got more advanced. "Our young target groups expect us to adapt to modern communications." Citygate had tested letting people input advertisements through text-messaging services. "Consider it—you only need the newspaper and the mobile phone to advertise." And when multi-media on mobile services (MMS) became cheap and available, the possibilities increased markedly. "Put MMS into the equation and you can snap a photo of your car and put it on the [web]site." Before joining Citygate, Jonason was manager of technology and new media for the Swedish newspaper association, Tidningsutgivarna, and then a management consultant with Ericsson's mobile business solutions group. "Everything we talked about at Ericsson in 2000 will come true—it is just a question of when. Just wait and see" (quoted in Campbell 2004: 9). Remember Saffo's 30-year rule?

The *Irish Times* generates $1.23 million a year in revenue by providing text messages of sports data and results to mobile devices (Crosby 2003). This is an example of a media group taking advantage of an innovative technology to make money. The key to the future may be each media group's willingness to embrace innovative technologies and ideas. Another occurred in the UK, when the online music distributor Napster signed a deal with News International, the publishing arm of Rupert Murdoch's News Corp., to distribute music through newspapers. It was the first time that free downloads were offered, via News International's mass-market daily tabloid, *The Sun*. As of July 2004 when the deal was finalized, *The Sun* had 10 million daily readers. Of those, about 4 million were aged 35 or under and considered prime consumers of online music. They had access to the 750,000 tracks available on Napster. As of mid-2004 the music industry was losing about $2.4 billion a year in sales because of Internet piracy, and the Napster-News International deal was seen as one way of dealing with this leakage (Burt 2004).

News publishers must prepare for a world where content is delivered to many digital devices. The web will remain important, but publishers must acknowledge that content will need a variety of distribution points. Non-traditional communication tools such as cell phone text messages are outflanking radio, television and print media because of their immediacy (interactivity is a major attraction) and the way that this forms a bond with users. Pedro Ramirez, editor of Spain's *El Mundo*, said the "thumb generation" posed the greatest new challenge to traditional media because of the capacity of cell phones to convey news, rumors and gossip via text messages. That challenge became evident after the train bombings in Madrid in Spain on 11 March 2004 that killed 191 people and injured more than 2,000 others, just three days before national elections. Nina Calarco, editor and publisher of southern Italy's *Gazzetta del Sud*, said information spread via cell phone messages contributed to the election of Socialist candidate Jose Luis Rodriguez Zapatero as Spanish prime minister. "The [Jose Maria] Aznar government in Spain was unseated by a shower of telephone text messages." These text messages contradicted the traditional print media that had repeated the government line that the train bombings were the result of Basque terrorism, Calarco said. The bombings have since been blamed on an alleged Moroccan-based Islamic extremist cell with possible links to al-Qaida. Ramirez said text messaging as a form of communication was difficult to control but easy to manipulate. "And whoever wants to insert himself into the chain can make an exponential effect during crises" (quoted in Redmont 2004).

A McKinsey analyst has suggested that commercial broadcasters in Europe could make money by linking television programs with text messages. SMS was an effective direct-marketing tool and can also increase ratings, analyst Jacques Bughin wrote. McKinsey studies found that adding SMS interactivity to certain shows improved viewer loyalty: "In some cases, the addition of SMS boosted the viewership of popular free-to-air television shows by up to 20 percent. Since advertising rates are directly linked to ratings, well-executed SMS-TV shows could at least preserve, and perhaps enhance, a broadcaster's bottom line." The 900 million text messages sent in the European market generated about $490 million for broadcasters, mobile operators and technology providers, or about 5 percent of the total SMS market in 2003, Bughin calculated. If the system's operational effectiveness were improved and if new shows, chat rooms, and shopping applications were developed, this market could easily be worth about $920 million by the end of 2005. Broadcasters could capture a third to a half of that sum (Bughin 2004: 1).

In October 2003 Nokia's executive vice-president, Anssi Vanjoki, told a World Association of Newspapers conference in Helsinki in Finland that cell phones and wireless provided excellent avenues for newspapers to reach younger demographics. "Older people use mobile phones for efficiency. For young people, it is more about excitement and fun. They develop incredible uses for these devices. They play with them." Vanjoki said this trend had serious implications for media businesses given that in the future most information would be digitized. He described cell phones and wireless devices as "media terminals" that were ideal for marketing static media such as newspapers (Vanjoki 2003).

Interestingly, the sensational media in the US appear most willing to experiment. In February 2004, Sprint and American Media launched daily wireless editions of *The National Enquirer* and *Weekly World News* on Sprint's PCS Vision service in a multi-media format for $3.95 a month. Wireless carriers are also developing next-generation networks that will deliver a form of television on cell phones. A company in Berkeley, California named Idetic launched MobiTV in November 2003. Subscribers must initially buy a contract for Sprint's PCS Vision data plan that provides unlimited access to the web and other online services. This cost $15 a month in 2004. MobiTV was then available for an extra $9.99 a month. Subscribers got 14 cable television channels, including ABC News Live, CNBC, the Discovery Channel, the Learning Channel, and MSNBC. Broadcast television screens at 30 frames a second, which generates the perception of movement. MobiTV broadcasts at one or two frames a second, more like a fast slide show than video (Landberg 2004). People who consider this slow should think back to the early days of the web, in about 1994, and then jump forward a decade to appreciate how quickly technology improves.

Impact on Journalists

What does this all mean for journalists and journalism? In some respects, little will change. Clear writing and provision of context and background will remain fundamental to the newspaper journalists' role. Clarity of expression also continues to be basic to the role of the broadcaster. But journalists will need to embrace technology and become more skilled with using it. In the light of the trends shaping journalism and the media, what kind of changes can we expect in the next decade? The rest of this chapter takes a big gulp of air—or maybe it should be something stronger—and makes some predictions.

The crystal ball, as mentioned earlier, is cloudy. But some things are likely. The print media will become increasingly visual. History shows us much about this process. If we inspect the front page of a newspaper from two generations ago and compare it with modern front pages, we notice the significant difference in design and the importance of visual impact. The newspaper is a reflection of society and the forces at work in that society. Visuals help people absorb information quickly and newspapers will work on ways to make their content more easily absorbed.

With more bandwidth, we will see lots more video on websites. Technological convergence has been happening for a generation. Eventually we may see the long-heralded merging of the personal computer with the television screen, so that much more of what we watch will be available via a hand-held computer screen, personal data assistant (PDA), or some form of device using electronic paper (e-paper). In short, the variety of devices for receiving and delivering news will increase. News is already a 24/7 operation, and that process will, if anything, develop in intensity.

Ben Compaine, a media researcher at Massachusetts Institute of Technology (MIT) in Boston, always points out that we should never underestimate the potential of a revolutionary innovation to change the way the media works (Compaine 2002). Think about how email and the web have radically changed the way journalists operate in many technologically-advanced nations in the last decade of the twentieth century. As discussed at the start of this chapter, the telephone similarly changed the newsgathering process from about the late 1920s and early 1930s. The cell phone is also affecting the way reporters operate in the early twenty-first century. The media industry still has not embraced the potential of cell phones as a delivery mechanism for news, but some bright person will find a way.

Increased Speed

News in the twenty-first century continues to be a fast-paced product requiring a similarly fast lifestyle. It was not always so. It took almost two months for news of the American Declaration of Independence, announced on 4 July 1776, to be reported in England. Details of Nelson's victory at Trafalgar on 21 October 1805 reached England a full 12 days later. If we jump forward two centuries we find digital information flashing onto computer screens across the world only seconds after it is released. Journalists need to appreciate and prepare for the ethical issues related to the dangers of increased speed. It is difficult to make wise ethical decisions on the spot. As photographers often say, you cannot make a good ethical decision at 1/250th of a second. The key is preparation: journalists need to be trained in good ethical decision making to deal with a changing, fast-paced world. Indeed, the need for continued training and education remains paramount, especially in the area of ethical decision making. Countries and companies without formalized codes of ethics for journalists are asking for trouble. Instruction in good ethical decision making needs to be part of every journalism program and every training course around the world.

Changing Deadlines and Attitudes

Convergence means a change in deadlines, and attitudes to those deadlines. Journalists will need to learn to allocate their time more wisely to deal with changing deadlines. The notion of a reporter working toward one sole deadline for the next day's paper will be as relevant as the need to crank one's car to start it—quaint but outdated. The converged journalist will need to be aware of all the deadlines in the building and allocate her or his most precious resource—time—to the most pressing. Editorial managers will similarly need to be flexible and willing to allocate resources to the most immediate news hole (this is a newspaper term for the space assigned for news). If journalists are to find time for reflection and considered judgment for "think" pieces, they will need to learn to juggle resources and be willing to ask for more time if they need it to complete those stories. Journalists will need to learn how better to manage their day (background reading may have to happen at home rather than in the office, for example). Time management will become a required and essential skill.

The larger the organization, the greater the need for forward planning and time management. The BBC offers an example of a huge news organization that spends vast resources on effective management of its editorial staff, often planning events months in advance. Teamwork and team building will become increasingly important. Editorial managers will need to allocate teams where necessary to do the job more effectively.

Communication skills will subsequently become even more vital. We may see the end of the lone wolf journalist who ignores colleagues and refuses to share information and contacts. In all cases these changing processes require a radical change in mindset.

More Technology

Reporters will need to become accustomed to using more and more sophisticated digital technology. Each year the Ifra staff at the Newsplex, based at the University of South Carolina at Columbia, assembles a collection of tools for better newsgathering known as NewsGear. Only the most innovative and practical of hundreds of pieces of hardware and software tested earn the "NewsGear" rating. For the past three years Newsplex staff have concentrated on technologies for multi-media reporting. The suite of tools must fit into a standard airline carry-on bag and cost less than $10,000. The 2004 collection cost about $8,765 in total; prices have declined every year since the project started. Journalists must be prepared to learn about new tools. This requires a mindset that is open to new ideas and ways of working. Chapter 3 described academic studies that pointed to conservative cultures in American newsrooms. Dinosaur institutions need help. Chapters 4 and 6 describe some of the tools available to journalists.

The Perils of Excess Information

We know it by many names: information overload, data smog, or information anxiety. Futurist Esther Dyson believes that digital technology potentially means that everything is knowable, and so the challenge is how to filter what we do not want to know, and ignore the information and people we do not want to hear from, as well as finding the best sources of good information. "Instead of finding, the challenge is filtering," she said. One consequence of this overflow of information for some journalists will be a move from newsgathering to news processing as a primary job responsibility. Perhaps in future more journalists will spend as much time each day, and possible more in some positions, editing and assembling the huge volume of news that arrives at a news organization, rather than gathering it. The need for quality editors, people who manage large volumes of information, will only increase. But the role of the journalist as provider of context and background will continue to be important.

Return of the Inverted Pyramid

For breaking news online, we will probably see a return to the inverted pyramid form of reportage as a way of getting information on the web quickly and effectively. Quality writing could become reserved for long-form pieces in the print medium, perhaps appearing only at weekends in quality newspapers or in magazines. Print journalism may break into two main forms: very short (breaking news online and news briefs on paper) and very long (considered, magazine-style writing and enterprise or investigative reporting, all of which are products of reflection and analysis). If journalism is to remain or become, as Joseph Pulitzer opined in 1902, one of the great intellectual professions, then the skills of analysis and synthesis will continue to be highly relevant. This form is most often likely to prevail at weekends (when people have the time to dedicate to long-form journalism) or in situations where professionals need deep content (such as in the pages of publications like *The Wall Street Journal*).

For other forms of news we may need to puncture one of the great myths of journalism, which is that it is primarily a literary form. Journalism is just as much about collecting and sorting good information as it is about expressing that information in a palatable form. If we describe a timeline for a typical news story, the amount of time spent gathering new information and processing existing information represents by far the largest segment of the line. But too many journalism schools and too many journalists think of reportage as a literary process. Literary journalism is needed for weekend newspapers and long-form articles, but during the week we should be giving our readers inverted-pyramid stories and bullet points. Synthesis may involve boiling down a huge amount of data and information into a few hundred words on a single subject, or providing a one-page summary of all the news and related events happening in a region or nation. Think of the way that the "What's News" section on the front page of *The Wall Street Journal* summarizes the key business and news events, and transfer that format to your local daily or website. We need to think in terms of audience needs and wants. Sometimes all that busy people seek is a synthesized form of news, a digest of events.

More Use of the Internet

It is a given that people's use of the Internet will rise. Let's look at some numbers for BBC television versus its fine website. As of early 2004, the main news site produced by BBC Interactive attracted an audience of about three million unique visitors a day. The highly-respected main evening news bulletin attracted an audience of about 10 million a day. But BBC Interactive was only five years old, while the television news needed half a century to build its audience. The online audience is growing while the television audience is declining relative to population growth. If we project a generation—about 30 years (remember Paul Saffo's 30-year rule from the first chapter)—the numbers suggest big changes are likely in news consumption patterns. Nielsen/NetRatings reported in 2004 that almost three in four households (204.3 million Americans) had Internet access from home. Broadband penetration in US homes had reached 51 percent, according to August 2004 data released by Nielsen. In towns like Lawrence, Kansas, the local media group (the World Company) reaches nine in 10 residents each day. How? It owns the local daily, the only cable news channel in town, the most popular group of websites, some niche-style magazines, and it is the monopoly provider of broadband and modem connections. This company makes good profits by selling multi-media advertising packages, as well as being involved in all forms of media. See chapter 5 for a case study of the company.

Online Journalism and the New Prime Time

One easy prediction is about the continued emergence and evolution of online journalism. Online is no longer new media. The Iraq war that started in March 2003 was a real tipping point for this form of journalism. Online provides a classic example of providing news when people want it. We have the evolution of a "new prime time"—people at work during office hours. Advertisers generally cannot reach people at work via traditional media such as television or newspapers, but they can reach people online. Websites cater for the work-based audience from 6 A.M. to 6 P.M., Monday to Friday. Proof of the existence of this new prime time can be seen in the logs of news websites that show that traffic drops by two thirds at weekends. In April 2004 the web research company Websense reported that 51 percent of employees surveyed said they spent only two hours a week on personal web surfing. But IT managers at the same companies estimated the figure at more than six hours a week, suggesting that people used the Internet at work more than they realized or were willing to admit.

Ben Estes, editor of the leading website chicagotribune.com, talks about the Tao of "floid" for online journalism. His "floid" was an acronym for content that was fresh, local, often updated, involving images and interactive, and aimed at the young demographic. The site's day started at 6 A.M. with content from *The Chicago Tribune* but quickly provided updates marked with time stamps and update tags. Local news was provided through breaking news with local angles: "Our franchise is not Iraq or Washington; it's local news," he said. The site concentrated on utilities—schools, weather, crime statistics, and traffic updates. In essence it was unique information that audiences could not get from the competitors. The site was updated as often as possible. Estes said he wanted the audience to perceive the site as fresh. Content varied depending on the time of day. The weekday morning concentrated on hard news. At noon the news was more utilitarian, such as advice about mortgages or how to get a better deal on a cell phone, while in the evening the focus was more on entertainment. "The stories people see at 9 A.M. are very different from the stories [they see] at 3 P.M. that day," Estes said. Images and a high level of interactivity were integral to the "floid" process. Readers loved photographs, he said, pointing out that they were usually the most viewed feature on the chicagotribune.com site. Interactive features such as polls and message boards were also very popular, Estes said, noting that with events that people cared about "these things take off like a rocket" (Estes 2003).

Content Still the Key

The essential part of all media is the content, rather than the distribution methods. Quality content attracts the audiences that media of all forms can sell to advertisers. (Another option is focusing on the other end of the market, but that is the subject of another book.) The key to the future of quality journalism around the world is the quality of the reporting, generated by intelligent, dedicated reporters and editors. It bears repeating that content was never king; there is more than enough content available in the world. Quality content reigns. That is why organizations such as *The New York Times* spend $250 million a year on producing unique content. For organizations unwilling to spend money to generate quality content, the future probably involves more repackaging of news, or a move to the bottom end of the market.

The tendency toward self-selected news, where younger audiences get information from Google or Yahoo or similar electronic sources, is a danger for democracy. This should be a worry to anyone who understands journalism's role in the democratic process, in providing people with the information with which they can make good decisions. One thing that media organizations need to do is to find ways to involve young readers and bring them back to the notion of reading newspapers and watching television news. Their actions may be a reaction against globalization or the alienation they feel toward media that ignore them. As an industry the media have much to do to find ways to build bridges with the 18–24 age demographic. Ways to reach that demographic include video, multimedia, and interactivity, and via forms of personalized media such as cell phones. As broadband penetration expands, we should expect to see more newspapers use video as a storytelling device. One of the beauties of convergent reporting is that print journalists who write for broadcast learn how to get to the point quickly and tell stories simply. The integration of text and image, via techniques like slide shows that combine photographs and a reporter's voice-over, offers an elegant way to do multi-media journalism.

Print and online media can also link to audiences via weblogs, better known as "blogs," and "moblogs" (a blog that uses a mobile phone). Newspapers like *The New York Times* have been experimenting with blogs since 2003. The issue with this latest tool is not technology but content. Academics and media managers need to figure out how journalism can give audiences something they cannot get elsewhere else. Blogging may come of age in the next decade. Major news events such as the Democratic and Republican national conventions in the lead-up to the US presidential election in 2004 were marked by the emergence of blogging as a new form of journalism. Reporters from many mainstream news organizations— including the Associated Press, CNN, *The Miami Herald*, *The Washington Post*, washingtonpost.com, MSNBC.com, CSMonitor.com, and Business Week Online—produced blogs and other online reports. Chapter 6 detailed how journalism students covered the Democratic convention in Boston in July 2004 and the Republican convention in New York in August 2004. Noted new media commentator Steve Outing, "If blogging isn't yet 'mainstream'—with the public or with newsroom executives— perhaps this [Democratic] convention will be what puts weblogs there" (Outing 2004c). In January 2005, the Pew Internet and American Life Project reported that 7 percent of the 120 million US adults who used the internet—about 8 million people—had said they had created a blog or web-based diary. Pew also said that by the end of 2004, 32 million Americans were blog readers (Rainie 2005). Technologies like blogs create communities and offer journalists a way to embrace technology not merely as a tool for reporting but as a way to bring people together.

Increased Need for Training

To use technology well requires appropriate training. The attitude to training in the newsroom must change. Journalism has become too complex a profession to pretend that training and education are not vital. It bears repeating that quality content generated by quality staff will be what differentiates great news organizations from mediocre ones. In the act of synthesizing information, journalists create knowledge, and with time these processes generate reputations for integrity and quality. Quality content will be the key to success, so news organizations will need to employ quality journalists to produce these intangibles. How do they acquire knowledgeable staff? The previous chapter noted the three main ways to generate knowledge: buy it (such as via a company take-over), rent it (hire consultants), or grow it (through research and development). Drives to reduce costs at media companies have meant that training budgets tend to be the first cut. Chapter 1 noted a 2002 national survey published by the American Society of Newspaper Editors (ASNE) in 2003 which pointed to lack of training as journalists' biggest source of job dissatisfaction, ahead of pay and benefits. "More than two thirds of journalists receive no regular training. Overall, news companies have not increased their training budgets since 1993. News executives acknowledge they should provide more training, but blame money and lack of time for their failure to do so" (Newton 2003: 9).

Early in 2004 the American Press Institute and ASNE started a project that could improve the situation. The Learning Newsroom is a three-year project designed to help newsrooms with their capacity to teach and learn the skills, knowledge and values they need to produce better journalism. Between 2004 and 2006, up to four newsrooms will be selected each year for one-year partnerships as model Learning Newsrooms. Management at the partner newspapers will work with the Learning Newsroom project director, Vickey Williams, to develop a learning plan to bring training consultants and organizational development experts into those newsrooms. Lessons gleaned from the experiences at the pilot newspapers will be shared with industry, Williams said. The Learning Newsroom came about after the Readership Institute at Northwestern University in Chicago identified workplace culture improvement as a top priority for improving readership and circulation (see chapter 3 for details). Experiences of the partner newsrooms will contribute to an industry-wide report suggesting ways to create better learning newsroom environments. The John S. and James L. Knight Foundation funded the Learning Newsroom as part of a $10 million training initiative. At the risk of sounding like a broken record, training must become a higher priority on managers' agendas.

This chapter will end with a continued call for quality journalism. The BBC's head of news, Richard Sambrook, has pointed out that when hearsay and second-hand information fight for people's attention, audiences respond to a trusted organization that bases its reporting on eyewitness coverage. In a crowded market, audiences recognize the power and integrity of original material. "What makes the BBC good at this is the breadth of our coverage and its depth. Breadth in that we have correspondents in 44 countries. We report for TV from over 140 countries and radio for over 160. Depth in that we are committed to reporting long-running issues over a period of time and to maintaining a strong corps of specialist correspondents." Sambrook said the BBC should aim to be distinctive and "do the things other news organizations won't" as well as provide a "backbone service for our audiences."

Since he took over in 2001, Sambrook said his strategy at BBC News and Current Affairs had moved from a concentration on distribution to an emphasis on original and distinctive journalism. "News broadcasting matters because reliable information and scrutiny of political developments are precious public goods. In an era of second-hand and third-hand information, spin, PR and lobbying, high standards are essential to democracy." Good journalism was vital to keeping powerful interests from abusing their position (Sambrook 2004). Convergence can combine the traditional values of journalism with the needs of audiences and markets. The key issue is the values of the people who run media organizations. The final chapter offers ways to implement convergence within this context.

9 Implementing Convergence in the Newsroom

This final chapter works from the assumption that the reader is thinking about introducing convergence, and suggests a process for how to do so. Because convergence is best led from the top it is important to have clear management objectives. The chapter begins by describing the factors common to successful convergence operations. It emphasizes the importance of communication and the power of shared values in driving successful change. And it returns to the importance of training, seeing it as an investment rather than a cost. The chapter ends by considering the factors that lead to resistance among journalists and other media workers, in the belief that an understanding of resistance will help managers appreciate what needs to be done for successful convergence to occur. The chapter looks in detail at ways to:

* Establish management priorities
* Recognize the importance of changing mindset
* Establish a convergence-friendly environment
* Recognize the power of communication
* Invest in training
* Resolve cultural issues

Change is a holistic process and should not be implemented as a linear or step-by-step formula. This chapter suggests a process for implementing convergence in a newsroom. Each organization will handle convergence in its own way, because each organization is unique. The process described here follows the themes outlined in various parts of this book: the need for changes in mindset, physical environment, and management attitudes. Convergence consultant Jimmy Gentry has proposed a convergence continuum thar outlines the factors that lead to "easy" or "difficult" convergence. Gentry's continuum is discussed in detail in chapter 2 (see figure 2.1), and is incorporated here in conjunction with the author's thoughts and opinions. This chapter proceeds from the assumption that the instigators have management buy-in.

The change process must begin with an awareness campaign driven by a thorough communication of values and intention. A training program based on investing in people and inculcating a multi-media awareness should be instigated as evidence of the organization's commitment to convergence and the people within the organization. Appropriate technologies for capturing and storing information, and ways to free reporters from their desks should be introduced as part of the training. Because modern newsrooms are part of the information age and produce knowledge (rather than processing data along linear industrial-age production lines), knowledge management processes should ideally be introduced at the same time. A key part of introducing convergence is the need to rearrange the structure of the newsroom and establish "symbols" of convergence, such as a multi-media desk. Allied with this is the need to publicize successes at all stages of the implementation process. Communication is vital at all stages, because communication is fundamental to any change process. Convergence pioneer Gil Thelen of the *Tampa Tribune* pointed out that his integrated newsroom was an extension of his work in team building and change management. "This really is a huge change process that's got a multi-media ribbon on it" (Thelen 2004a). Part of the communication process involves understanding the cultural differences between different media, and resolving the stereotyping based on ignorance of how other media operate. We now look at each of these parts of the process.

Establish Management Priorities

Because of the way that media companies are structured, it is implicit that any company's decision to adopt convergence journalism will come from management. But any application needs to be driven by the editorial team. As has been shown in chapters 2 and 3, the introduction of convergence is not about saving money. It is a way to reach more of the scattered audiences produced by media fragmentation. It ensures media companies' viability because those companies can reach more of the fragmented audiences. Tampa's Gil Thelen said the long-term benefit of convergence was the capacity to respond to a changing market place. "We're building a news organization that has the capacity to respond adroitly" (Thelen 2004a). Convergence offers a way to do more complete journalism, in the sense of telling stories in the most appropriate medium for a specific news event. It is a way to revitalize some senior journalists' careers, and it certainly gives journalists more marketable skills. In essence, convergence gives editorial managers and journalists a way to improve audience reach (and potential revenues) while at the same time doing better journalism. For all this to happen, direction must come from committed managers who believe in this dual potential. This means that managers must have clear priorities and a roadmap for introducing convergence. They must demonstrate the advantages of convergence and communicate the reasons why it must happen. Ifra's Kerry Northrup said a boardroom mandate was needed for any "serious convergence conversion" of a media operation. "But grassroots adoption is the only successful approach to implementation" (Northrup 2004d).

Paul Cheung, chief editor of the *Ming Pao* daily, part of the Ming Pao Group in Hong Kong, emphasized the need for clear direction from managers. Though the decision to embrace multi-media came from the group's managers, change was "editorially driven," he said. "We knew we could not stay where we were. We had to move on." Cheung employed a form of "helicopter vision" to get an overview of the situation. "An editorial manager must get above the daily process to see where they are going." He had a clear sense of purpose; his aim, he said, was to create a new form of journalism. "We are trying to develop a new breed of multi-media journalist. It takes time" (Cheung 2001). Tampa's Gil Thelen said his role was to push convergence constantly and then demonstrate how the collective product was better for the community and ultimately for journalism. "Repetition is hugely important in setting up ways of keeping track of what you have done and where you are, and reminding people constantly of it" (Thelen 2004a). Northrup said the management complexity of handling stories in multiple media for integrated distribution was significantly greater than what most newsroom managers had experienced. "It is relatively easy to find and train journalists to work cross-media, compared with preparing their editors to work nonlinear in the same environment" (Northrup 2004d).

Recognize the Importance of Changing Mindset

Andrew Nachison, director of the Media Center at the American Press Institute and a frequent speaker and consultant on convergence, said inculcating the appropriate mindset was a key management process. Among the significant operational and strategic challenges involved in morphing from a single-platform operation to a multiple media system, one of the most important was the need to manage the change process itself, he said, and to help everybody in the organization commit to the purpose of a multi-media news company. The challenge was not limited to the newsroom, Nachison said. "Sales, marketing and even support staff need to come to grips with what the enterprise is all about. All media have developed standards, traditions and even language, and convergence at times involves abandoning decades of learned behavior and attitudes."

Nachison pointed out that the 24-hour news cycle was nothing new. Wire service reporters had been working comfortably with it for years. "The information is the driver, not the delivery platform. That's the lesson that can be most difficult to learn: that the newspaper, TV station or website you once thought you worked for exclusively is no longer your priority. You are no longer exclusive to any platform. Your priority is getting the information out on whatever platform it should be on." Dealing with this new world view took time, training, executive leadership and money, he said (Nachison 2001a). Three years later Nachison concluded that the greatest challenges in initiating convergence were the issues of leadership and how to pay for the changes. "Where there is failure, I would say it's primarily a leadership failure—an execution or tactical failure, or maybe sometimes an investment failure. If you don't have the leadership and the financial investment to make it happen, you're going to fail. The fact you have employees who resist does not mean convergence cannot work. If organizations genuinely want to change they will do what they need to do to change" (Nachison 2004).

Northrup said convergence involved a fundamental shift in strategy: evolving from a content production and distribution industry to an information service industry. When training people at the Newsplex, the "primary goal" was "developing the multi-media mindset," Northrup said. "Despite all the technology, this place is not about technology. It is about the thinking processes for the new world of journalism. It is about mindset. People do not come here to learn software. This place is about thinking [and] about multi-media storytelling" (Quinn 2004). Northrup said that in an information economy, news was a commodity and value was placed on services that informed, explained, and evaluated. "Convenience is the primary attribute of a service. Convenience in a news and information service dictates that it operate in whatever medium a customer favors." Chapter 2 discusses the issue of convenience. A key part of Northrup's process involved a changed mindset among journalists. This necessitated a fundamental shift in the newsroom: focusing more on content and information management than on traditional production management. "This is essential to support the service-oriented multiple-media company" (Northrup 2004d).

When accepting the chairmanship of the Newspaper Association of America in April 2004, Gregg K. Jones urged newspaper executives to innovate rather than remain content with the acknowledged strengths of the industry. Jones is co-publisher of *The Greeneville Sun* in Tennessee and president of Greeneville-based Jones Media. "I firmly believe," he said, "that as an industry we are at a critical juncture. We have two choices. We can accept our future. Or we can invent it." Jones predicted that newspaper companies would become information companies that included newspapers, specialty publications, the web, and wireless. "It's not a departure from what we do," he said. "It's a logical and natural evolution." He also suggested that a multi-media future was "not just for the big papers" (quoted in Anonymous, *The Greeneville Sun*, 2004: A1). John Haile, editor of the *Orlando Sentinel* when it evolved into multi-media company, said a vision for a new kind of company—with journalism still at its core—began to develop in the company's strategic and operating plans. Over time it produced wide discussion among managers. "Because of the [company's] early work with AOL and the Internet and, then as the *Sentinel* prepared to launch a high-profile test of interactive television, it was clear that the traditional newspaper must become a multi-media information company. That required a whole new set of skills and a different way of thinking about news, information and advertising. It also meant that the newspaper's content had to assume different forms" (2003: 5).

Ifra's Northrup said that not all journalists in a newsroom had to be multi-skilled—"perhaps only a minority." But all had to be "multi-media minded" and all journalists in a multiple-media news organization had to understand the strengths, weaknesses, and capabilities of all formats through which their stories could be presented "if they are to be able to create the most effective telling of those stories" (Northrup 2004d). Chapter 4 discusses the strengths and weaknesses of each media. Haile also saw the potential financial benefits of convergence. He repeatedly argued that audience fragmentation produced the greatest risk to a strong journalistic enterprise. Revenue streams would begin to dry up, meaning fewer resources for the newsroom. "So, there was intense interest in finding ways to move with the audience as technology changed the way people get their news and information. In doing that, the newspaper could provide strong journalism while building new relationships that ultimately could bring in additional financial resources" (2003: 10).

Establish a Convergence-Friendly Environment

Once a decision has been made to implement convergence, editorial managers must establish an environment that allows it to happen. They must establish processes for changing attitudes and mindset. Gentry said one way to align journalists was to concentrate on journalism's core values. During the introduction of convergence at the *Orlando Sentinel,* then editor John Haile and then managing editor Jane Healy held regular "news values" discussions with small groups of staff. This involved taking people out of the office for a day and making time for discussions about what constituted good journalism. Some staff were concerned that convergence would lead to a change in the newspaper's values. "We said they wouldn't," Healy said. "We said we want to put our [the newspaper's] values on TV." Haile emphasized basic journalism values. "We made a point to emphasize that there were fundamental values that will define us no matter what medium we are in. These values set us apart from the competition. They are our competitive advantage" (quoted in Gentry 1999: 7). Managers at the World Company in Lawrence, Kansas, similarly organized seminars and workshops for small groups of staff. They brought together people from different areas of the company in these workshops, where people were able to express their concerns. The most powerful way to establish a convergence-friendly environment is through communication and by using creative strategies to get people from different cultures to work together.

Recognize the Power of Communication

For the change process to work smoothly, people need to be told from the outset why change is being introduced. Convergence consultant Jimmy Gentry said that for a change effort to be effective, affected individuals and groups "had to understand why the change must occur." Gentry has a background in management as well as journalism, and advised the *Orlando Sentinel* on its convergence moves. "Leadership can never communicate too much in a change effort" (Gentry 2000). *Tampa Tribune* publisher Gil Thelen said it was vital to repeat the message as part of the communication process. "You've got to say the same thing hundreds of times in dozens of venues before it reaches all levels of the organization. About the time you're getting bored hearing yourself talk, you're just beginning to really communicate effectively" (Thelen 2000c). Four years later, Thelen re-emphasized the importance of communication. "It's a key factor. It must be maintained to continue the process" (Thelen 2004a). Regular newsletters offer a good way to keep people informed. These could be provided in paper form or via an organization's intranet, and could involve publishing daily lists of the most popular online stories and publicizing convergence achievements.

Communication Reduces Uncertainty

Humans dislike disruptions and need to see the reasons *for* change and the reasons *to* change. Reasons *for* change are described in the first two chapters. In terms of the reasons *to* change, editorial managers need to show journalists how convergence will affect their lives, and how it could make their work more interesting. Convergence offers journalists ways to be more powerful advocates of the media's First Amendment responsibilities. Reason to change could be introduced along the lines of improving each individual's professional qualifications. Paul Cheung, chief editor of *Ming Pao* in Hong Kong, said his staff accepted the concept of multi-skilling because he told them it would enhance their employment prospects. To manage the change of mindset, Cheung held weekly meetings with senior editorial managers, who in turn communicated with their staff. Between June and December 1999 he used meetings to introduce new ideas. Cheung argued in the meetings that integrated journalism was a chance for *Ming Pao* journalists to upgrade their skills. "We are helping them to improve their professional standing."

Corporate culture played a key role in *Ming Pao* journalists' acceptance of convergence, and the process took six months to percolate through the newsroom. Chinese respect for authority also helped the process, in the sense that people in that society were more likely to follow the instructions of authority figures. But the key, Cheung said, was getting people involved and understanding the reasons for the convergence process. "My staff are not the highest paid in Hong Kong but they believe in what they're doing" (Cheung 2001). Ifra's Northrup said the fastest and most painless way to improve communication and cooperation was by co-locating staff from different media and different organizations. Chapter 5 discusses how the World Company in Lawrence, Kansas, has co-located reporters from different media who cover the same beats. Northrup said key editorial managers from converging news organizations especially benefited from sitting and working together on a "super-desk" that served as a "central and concentrated news-handling resource for the entire cross-media enterprise" (Northrup 2004d).

Gentry said management and staff cooperation at the *Orlando Sentinel* was the result of widespread consultation. Then editor John Haile told staff how he and editorial managers wanted to "change the company." Before the partnership with local cable Channel 13, the newspaper formed six committees. Two groups looked at content issues; another assessed the impact on staff; a fourth examined technology issues; another considered work flow issues; and the sixth assessed training needs. Then managing editor Jane Healy said almost 100 staff members became involved and "this allowed us to get buy-in by most of the staff. They had enormous input into how we would do things." Haile said newsroom leaders generated ideas and felt ownership of the process. "The leaders in the newsroom got out in front and said 'this is vitally important in reaching more people and we should take advantage of the opportunities to experiment.' They jumped in" (quoted in Gentry 1999: 7).

Mission statements offer a way to align staff. Gloria Brown Anderson, vice president for international and editorial development at *The New York Times*, described how a series of meetings held in 1992 produced a "keystone" in 13 words—a fundamental statement of the organization's mission: "Editorial excellence and independence are essential to our profitability and profit sustains them." Anderson said the wall between the business and news sides of the paper was maintained but the bonds were stronger because agreement had been reached "as to where we can appropriately approach the line and where we cannot." Managers later produced—in another 13 words—the company's core purpose: "To enhance society by creating, collecting and distributing high-quality news, information and entertainment." Anderson said these statements had enabled *The New York Times* to pursue its goals and objectives in a more coherent fashion, and added that a commitment to editorial excellence and independence had put the paper's editors at the "heart" of the newspaper's operation. The process she described had served to "emphasize the primacy of the editorial content" (Anderson 2001).

Gentry conducted a SWAT (strengths, weaknesses, assets, and threats) analysis when introducing a convergence curriculum not long after taking over as dean of the University of Kansas's school of journalism and mass communications. Part of this involved unearthing the faculty's values. Through a series of processes—"process is key"—Gentry said faculty agreed upon a student-centered values statement. One of Gentry's first acts was to create values committees with respected senior academics as leaders. Concentrating on values helped build foundations, he said, which got the heavy flywheel of progress moving. Though the wheel moved slowly at first, it gained momentum. "The values statement is what makes it happen." Gentry also encouraged people to give up the past and its issues and look to the future (Gentry and Musser 2004).

Newsplex training director Martha Stone noted that in the convergence environment "it is all about working together and communicating." How could managers build bridges instead of putting up walls, she asked. "It is not a secret when making a change like when going from mono-media to multi-media that it is always the people issues that are the most difficult." Stone said Newsplex staff emphasized the importance of people issues during training programs. "We take special care to make sure that the people who will re-train their personnel are very aware of how important people issues are and how they can bring back strategies to their companies to address those issues" (quoted in Roper 2003a: 30). Stone was referring to an aspect of classic innovation theory that showed that in deciding whether to adopt an innovation people respond to subjective evaluations by others like themselves.

Innovation theorist Everett Rogers said people modeled their adoption behavior on the attitudes of opinion leaders and noted that the success or failure of diffusion programs often rested on the role of these opinion leaders (1995: 89). Rogers wrote that the most striking characteristic of opinion leaders was their unique and influential position in a system's communication structure. "They are at the center of interpersonal communication networks. A communication network consists of interconnected individuals who are linked by patterned flows of information. The opinion leader's interpersonal networks allow him or her to serve as a social model whose innovative behavior is imitated by other members of the system" (Rogers 1995: 27). Northrup succinctly summarized Rogers's theories: "To generate grassroots adoption [of convergence] in the newsroom seed the operation with change agents trained in the skills and mindset of multiple media" (Northrup 2004d).

In a wide-ranging study of convergence at four newsrooms, researcher Jane Singer noted that interpersonal communication channels were of "primary importance" in the diffusion of convergence within those newsrooms. She also noted that in their descriptions of convergence, journalists used words usually reserved for describing marriage, writing that: "Relationships develop when journalists sit 'elbow to elbow' one broadcast news manager said. 'Proximity breeds collegiality, not contempt.' Several journalists compared convergence to a marriage; in Tampa, the negotiations prior to convergence were commonly referred to as the 'pre-nups.' Journalists emphasized the need for 'commitment and trust' among participants—and the time to develop it." Singer quoted a television reporter from among the people she studied who said,

"You have to work at it, understand each other's idiosyncrasies, go from there." Singer described convergence as "a marriage of convenience," and concluded that cultural clashes remained a major stumbling block to convergence which "may well be a hallmark of the process in every newsroom." She also reported that many journalists said they had gained respect for people in other areas and functions of the newsroom after working with them (Singer 2003a). This introduces us to the notion of cultural difference and how to deal with it.

Ball State University researcher Vince Filak has argued that journalism is socially constructed, noting that socialization started early in a journalist's career, often with selection of a major at university. Journalists formed groups or tribes and the "in group" tended to be suspicious of the "out group." This produced inter-group bias. One way to deal with this bias was through what Filak called the use of supra-ordinate tasks such as forcing people to work together for the greater good. For organizations hoping to introduce convergence this would be one way to resolve inter-group problems. "Contact is not enough," he said. "Getting to know a person beyond the task is important." Supra-ordinate tasks such as requiring student teams to learn new skills forced them to find others who had the needed skills. The shift of identity focus dissolved bias (Filak 2003). Media managers needed to be aware of inter-group conflict, Filak said. Even if a plan to converge newsrooms were well thought out and beneficial to all concerned, the perception that it came from "out group" members could damage its standing. Members of all groups needed to be involved in creating a convergence plan; otherwise those who were left out were likely to reject the plan, he said. Researchers in psychology had found several ways to decrease inter-group bias. "They have found that certain tasks and goals diminish bias among previously competitive groups." Solutions found by other researchers should be applicable to the newsroom setting to improve the transition from many newsrooms to one, Filak said (2004: 23). This supports the work of Gentry and Singer, who noted the power of having people work together on major projects as a way to introduce convergence.

Gentry and Musser at the University of Kansas's school of journalism said issues of cultural differences and stereotypes could be resolved through agreement on values. Gentry said he always started with direct discussion about values. "If you put the TV team and the newspaper team in different corners of the room, get them to write down their values, and then put the results on the whiteboard, almost all the time you find they are the same: truth, honesty, integrity, service to community." Gentry said the key was revealing those values to the other side. Discovering how the other group worked generated another set of appreciations. Television news was "damn hard to do," Musser said. Musser, who had a print background, spent the summer of 2001 at WGN-TV in Chicago on a fellowship to learn about television. Each year the Radio and Television News Directors Association (RTDNA) offers fellowships for broadcast academics to refresh their skills by working the summer in industry. Musser was the only newspaper person on the program. "Work a day with the other medium is good; work a month is even better," Musser said. He was hinting at an American Indian saying that before one condemns another, one should "walk a mile in their moccasins" to understand how they live their life. If they did, journalists in both media would realize that some of the preconceptions and stereotypes about other media were wrong. Students at the University of Kansas journalism program were required to work in teams to produce a print story and a television story about a specific topic, such as a government policy. Gentry said convergence worked when an organization had people such as Musser who were willing to embrace new ideas. "It's easier if respected members of faculty are willing to find out about new things" (Gentry and Musser 2004). They were referring to the powerful role of change agents that innovation academic Everett Rogers has described (1995: 27).

Keith Wheeler, associate managing editor for broadcast and online at the *Orlando Sentinel* and a former television news director, said the main cultural differences between print and broadcast journalists was the latter's sense of immediacy—being able to turn stories around quickly—and their work practices. People in a television newsroom worked toward several deadlines in a day, while print journalists usually had one. Newspaper reporters could interview people by telephone, while television journalists needed to take a camera out to get images and sound bites. "When doing a story for television you have to get up and go out." Both types of reporters were giving the audience as much as they could in a limited amount of time. "TV and radio are there to whet people's appetites. Print people can offer more detail." Both types of journalism involved asking the right questions and doing thorough research. "It's just that television has less time to do a package." Wheeler said print and broadcast journalists had more in common than opposition, and part of his role was to get the editorial staff to appreciate the similarities. Convergence was about making news available in all forms to reach as many people as possible. "You make the information work for people; one story serving more than one purpose" (Wheeler 2004b). Ifra's Northrup noted the value of mixing the tempos of newspaper and television in an integrated newsroom, in terms of energy level and pace. The glossary at the end of this book is designed to help journalists and students understand the key vocabulary of each medium.

Invest in Training

For convergence to succeed, an investment in training is vital to help journalists perform to their potential. This is itself a change of mindset, given the paucity of training in American newsrooms. Tribune Company president Jack Fuller has noted the problem: "The editorial departments of newspapers are notoriously poor in offering training" (1996: 183). Lack of training can cause problems when technology is involved. Diffusion of innovation theorist Everett Rogers noted the "tool nature" of technologies associated with the communication media. As with all tools, people adapted and used them in different ways, and this produced a high degree of modification from the original purpose of that technology. Rogers used the term "re-invention" to describe this process and noted that computer-based technologies were frequently characterized by a "high degree of re-invention." The negative side of this re-invention process was a tendency to waste time through poor use of technology. And individuals who struggled with technology tended to criticize the technology when discussing it with colleagues. Sometimes these water-cooler conversations were more powerful than the advocacy of opinion leaders. One should never underestimate the power of interpersonal channels to boost or hinder the adoption of an innovation (Rogers 1986: 121).

Training and communication remain the most efficient ways to remove the feelings of inadequacy that individuals sometimes feel because of their ignorance of technology. Training also provides an opportunity to spread the word about the benefits of convergence, and gives journalists an outlet for expressing their fears and concerns about it. It is easier to deal with these issues in the classroom, where a knowledgeable instructor can allay people's concerns and demonstrate the benefits of new forms of journalism. Fear is usually the biggest barrier to learning a new technology skill. Training managers need to be aware of people's fears and establish processes that make it easy for people to enroll in classes. Time is often an issue for busy journalists, so programs need to be flexible. And managers need to demonstrate their commitment to training by making time available, requiring busy journalists to attend, and leading by example by attending courses themselves.

Training need not be expensive. One effective approach is to recruit journalists with television experience into the newspaper newsroom. They become mentors and trainers. In February 2000 Ming Pao hired five former television journalists to help train their print staff. These included William Fung and Martin Lee, formerly of TVB, Hong Kong's highest-rating commercial station. Fung had been the news manager and Lee the assignment editor. Fung and Lee attended all main meetings to keep abreast of convergence planning and to familiarize themselves with newspaper production processes. Cheung said they brought much needed television experience and knowledge with them. "And they are excellent trainers." Fung and Lee organized lunch-time courses and were available for consultation in the newsroom. Lee said many print journalists in Hong Kong were initially scared of television. "Mostly it was because they knew nothing about the technology. But they soon learned to cope and after a while became proficient." By the end of 2000 all 40 reporters on Ming Pao's local news desk were working as multi-media journalists. They carried relatively inexpensive Sony digital video cameras, bringing their field tapes to in-house editors who cut the tape into stories suitable for broadcast and the web.

The *Orlando Sentinel* similarly hired seven television journalists, producers, and production staff in 1997 with the intention that print staff would learn from their new colleagues. The *Sentinel*'s Keith Wheeler said television journalists had a sense of immediacy. The paper's broadcast producers helped print reporters appreciate the time pressures that broadcasters worked under and taught them the skills they needed for voice-overs and talk-backs. The latter are question and answer segments where the television anchor in the studio talked with a print reporter in the *Sentinel*'s newsroom. This training also helped break down cultural differences. Journalists working in multi-media needed to know how to write for television, and how to conduct talk-backs. A good multi-media coordinator knew how all aspects of the media worked and provided lots of training and support for journalists, Wheeler said. "I'm nurturing these people. I don't make them do things they don't want to do." Wheeler said he had to earn reporters' trust: "When I arrived in 1997 my biggest job was PR" (Wheeler 2004b).

The Chicago Tribune also offered extensive training, using in-house photo-editors and photographers. Mark Hinojosa, assistant managing editor for electronic news, said the newspaper would not train reporters to be still photographers or expect them to replace still photographers. The only exception was foreign correspondents. "We give them [foreign correspondents] enough training to make a decent photograph to go along with their stories." Professional television camera people shot video for the television partners WGN and CLTV. "We create our packages for television, and we use professional television photographers and editors because of the size of the audience compared with the newspaper." With broadband becoming ubiquitous in major cities, people expected video to be of broadcast television quality. "That's a model that people understand and appreciate. Video on the web as we make the transition [to broadband] will have the look and feel of television. You can create that but it takes a lot of work, and a special talent. Just as being a still photographer requires a special talent" (Hinojosa 2004).

Resolve Cultural Issues

The Media Center's Andrew Nachison said television and print news divisions had different cultures but this could be resolved through a commitment to education and communication (Nachison 2001a). Ifra's Northrup advocated training the entire newsroom in convergence, not individual journalists and editors. "It's a process, not a skill set" Northrup 2004d). Anthony Moor was new media editor for the *Democrat and Chronicle* in Rochester in New York State before becoming editor of orlandosentinel.com. He said he had to educate staff in Rochester when he helped develop a partnership between the paper and two television stations. His newspaper was surprised that one of the stations agreed to provide its online weather page, Moor said. "From the publisher on down they didn't realize what television would do as a partner." The newspaper also did not realize the power it had to attract TV stations into a valuable partnership, he said (quoted in Dotinga 2003). Northrup emphasized that the newspaper's greatest asset in a cross-media partnership was the size of its reporting assets, while television's best asset was its "emotional connection" to news consumers. "A properly structured newspaper-TV interaction benefits both players, though in different ways. TV gets depth and reach [from the large number of reporters in the newspaper newsroom]. The newspaper gets immediacy" (Northrup 2004d).

Initiate Knowledge Management Processes

The final section of chapter 7 covers this subject in detail, so this chapter will not dwell on it. One of the aims of knowledge management is to foster collaboration among staff, which helps to dissolve the differences that cultural stereotypes generate. Collaboration also builds trust. Appoint a knowledge management specialist and give them high rank and an office near the center of the newsroom. Assign people to work in teams and elicit feedback on what works and what does not. Show examples of successful team efforts. One way to encourage collaboration is to set up environments where journalists congregate. These "magnet locations" become places where people get to know each other. Saf Fahim, the architect who designed the Newsplex building, has established open plan areas in newsrooms where staff can socialize and bond. This in turn fosters trust. People were more likely to form groups to take on projects if an atmosphere of trust prevailed. Integrating staff from different backgrounds was another way to establish an environment in which collaboration might evolve (Fahim 2004). The *Maeil Business Daily* in Seoul, South Korea opened a fitness club. Dae-Whan Chang, president of the Maeil Business Group, said journalists spent time in the club socializing with colleagues, often discussing work. "Often journalists come to work early to talk about a project they are working on" (Chang 2001). Northrup suggested that print newspapers could evolve into a premium news product because print remained a preferred medium among consumers for knowledge transfer (Northrup 2004c). See the discussion about "affordances" in the section on each medium's strengths and weaknesses in chapter 4 for more details.

Provide Incentives and Rewards

Journalists are not known for playing well with others. One way to change this kind of culture is through incentives. Journalists at *Ming Pao* in Hong Kong received yearly performance appraisals and chief editor Paul Cheung awarded higher pay raises to journalists who showed a genuine commitment to integration (Cheung 2001). Until 2004 Janet Weaver was executive editor of the *Sarasota Herald-Tribune*, which has partnered with cable television since 1995. Weaver said reporters' contributions to a converged newsroom were part of their performance reviews and development plans. High performers were often rewarded through merit raises and promotions (Shearer 2003). Tampa's Gil Thelen similarly said journalists who undertook multi-media reporting got a larger pay raise than other reporters (Thelen 2004a). The Tribune Company also provided incentives via recognition during annual performance reviews. *The Chicago Tribune*'s Mark Hinojosa said participation in convergence was an item in journalists' annual evaluations (Hinojosa 2004). Northrup has noted the importance of measuring convergence accomplishments. "The newsroom staff needs a tangible indication that progress is being made. For instance, count the number of stories handled in a converged way; give recognition to the most innovatively cross-media news effort; track news consumer response to particular stories by combining readership ratings, web page hits, and the number of emails/letters/phone calls received" (Northrup 2004d). Thelen said convergence needed measurable goals and "mileage markers" (Thelen 2000c). Successes are displayed in the conference room at the *Tampa Tribune* where each day's convergence meeting is held.

Establish Symbols of Convergence

Symbols resonate with people. The multi-media desks that have arisen at newspaper offices represent physical and metaphorical reminders of the commitment to change. Tampa's Gil Thelen said the multi-media desk was a continual reminder of the changes taking place in the Tampa News Center. The continuous or advanced news desks at major American dailies have become recognized as emblems of change, as well as the place where breaking news is assessed and managed. Northrup said "super-desks" were some of the first examples of a newsroom's intention to embed cross-media news handling. "Continuous news desks are an evolution, serving a liaison function between media." To manage the integration of media, editors had to cross boundaries between content formats. "Creating a convergence or multi-media editor with responsibilities and authority over more than one medium is a common start and can eventually lead to, for instance, a sports editor who manages all sports coverage in print, online, and on the air" (Northrup 2004d).

The World Company in Lawrence, Kansas, established a multi-media desk in 2001. Convergence director Rob Curley described the dedicated multi-media newsroom in the building as "our ode to convergence." The flavor of convergence there was different from bigger newspapers, and the newsroom was divided not by media, but by beat. "The cop reporters for TV and print sit next to each other. So do the education reporters." The newsroom also regularly produced five to seven local television programs, featuring subjects like home and garden and local music. "Some of the [television] people who produce the local origination programs sit with the [newspaper] features staff" (Curley 2004a). He was relating to another key aspect of convergence, which is the need to change the structure of the newsroom to generate synergy among journalists.

Figure 12. The multi-media desk at the *Lawrence Journal-World* with its polished wooden floors. Photo: Stephen Quinn.

Move the Furniture

Aim to avoid having tribes or clans of similarly focused reporters sit together. Move the desks so that reporters and copyeditors work together. This has the added advantage of breaking down more barriers. Roster copyeditors to arrive before reporters leave, to work on timeless copy or features, because it is important to erase the stereotypes between copyeditors and reporters. Look for ways to get members of each group to talk to one another. Re-introduce the Maestro concept (where a team of a reporter, editor, photographer and graphic designer work on one story). Put reporters on the copyediting desk to learn another skill. Along the way they begin to appreciate the editors' roles and responsibilities. Similarly, roster copyeditors to work as reporters so they can understand the pressures that reporters face. Sometimes it is not possible to put people together. Ten journalists from the *Manchester Evening News* in the UK constituted the first group to be trained at the Newsplex. After the training they reported improved teamwork, and requested structural changes. Because of space restraints the newspaper's news desk and web operations remained physically apart, but a Manchester Online journalist was given a place on the news desk to improve coordination between the two media (Veseling 2003: 103). Managers accepted that communication was vital even if it was not possible to put people together physically. Northrup said it was vital to design the newsroom to support the convergent mission and news flow. He suggested installing a "newswall"—a screen perhaps eight by six feet where video feeds, page displays, and the online site could be displayed—as a focus point for editorial interaction. People could comment on live pages being constructed, sometimes suggesting better headlines or cutlines, or noting errors. Managers should also supply reporters with cell phones and wireless computers to enhance their flexibility, and provide locations for talkbacks and television recording (Northrup 2004d).

Establish Common Information Databases and Use Technology Wisely

Journalists deal with information as their raw material. For convergence to work fully, journalists need access to all the information that enters a newsroom. The best way to store this information is in one database or in a series of interconnected databases. Technologies such as XML (eXtensible Markup Language) which involve tags that provide easy access to the information in the database are useful. XML enables journalists to access and reformat content into any media form. Any technology needs to be simple to use. The journalist is then able to concentrate on shaping the information for the most effective medium to tell the story, and not waste time changing formats. Any technology system that requires people to re-key content extensively from one digital form to another is literally a waste of time and energy and should be replaced. Over the first decade of the twenty-first century we should see some new database tools emerge for easy re-purposing of content. Some of the likely leaders are NewsGate (designed by CCI in Denmark) and Control Tower (by Proxim-IT in the UK).

Associated Newspapers is one of the biggest media groups in the UK and publishes major dailies such as the *Daily Mail*. Allan Marshall, the company's chief operations and technology director, said Associated Newspaper reporters were experimenting with the use of Blackberries for newsgathering. "We are also investigating the value of 3G [third-generation] in terms of transmission of pictures." Wireless networks were also becoming more common, he said, "and we will soon have devices that can switch between public and private networks, so for example we will see wireless access in council offices or major companies that reporters can utilize" (Marshall 2004). Northrup said it was important to put the story and the content management system that supported it at the center of the technological infrastructure, rather than traditional production processes. "Databases become a core corporate asset as they are in other information-based companies." He applied one simple measure of a converged organization's core system: "Is there a place to record, develop, communicate, and track a story idea? And when the story is done, is there anything to show for it other than what is finally published and aired?" (Northrup 2004d). He was alluding to the importance of knowledge management systems in the newsroom and throughout the organization.

Listen to Dissenting Voices

Any manager interested in implementing convergence should consider why people might oppose it, to understand the fears behind the dissent. This section briefly discusses dissenting voices and looks at why people oppose convergence. Researchers at the Mudia consortium in western Europe have compiled a list of the factors leading to resistance to convergence, and the author has added to it. (Mudia is an acronym coined from the phrase "multi-media in a digital information age.") Two major resistance factors were the fear that convergence would force journalists to work harder for no more pay, and convergence would lead to a dilution of voices in the community. Linda Foley, the president of America's Newspaper Guild, said that the guild had no problem with the concept of convergence except when it forced workers to do more work for the same pay. At Tampa's News Center, print reporters had had to adjust to meeting deadlines 24/7 and to rewrite stories, she said. "It puts an incredible amount of stress on the reporters themselves, because they're expected to serve many masters," Foley said. "It's very difficult to produce news and do stories for various media" (quoted in Glaser 2004a). As of mid-2004, very few media organizations around the world required journalists to be multi-skilled Inspector Gadgets, and in many cases the process was voluntary. This may change with time as managers decide to impose multi-media requirements on new hires, and as we see the arrival of keen graduates from university programs teaching this form of journalism. Northrup noted that not all journalists would make the transition to working for converged media companies, or to thinking in terms of multiple media rather than their specialization. "Media managers should be prepared for how they will deal with those who cannot adapt" (Northrup 2004d).

The issue of diversity of voices can be linked with market size. *Tampa Tribune* publisher Gil Thelen said enough diversity and competition was possible in markets of size 60 or larger. (The US broadcast environment is divided into 210 markets ranging from the largest, New York City, through to what is believed to be the smallest, in Glendive, Montana.) Multi-media journalism would not silence diverse voices in 150 of the 210 markets, Thelen said. "In smaller markets it is a relevant question," he said. "A lot has to do with the motives, values and principles of the owners," he said, noting that it was difficult to legislate against people's motives and business practices (Thelen 2004b).

Other factors that produced opposition to convergence included the expectation that the introduction of convergence would be a painful experience, because of a fear of the unknown. The likelihood of minimal training also concerned journalists, based again on past experience. The lack of a business model can also interfere with the successful introduction of convergence. Journalists have expressed concern about how companies could pay for convergence. These are all legitimate concerns which must be resolved. It is worth repeating that communication helps smooth the change process. It comes down to the quality of management and their ability to communicate. Training must happen and it must be perceived as a benefit rather than a cost. Convergence is not a way to save money, especially in the early stages. Viable business plans must be created and negotiated (see chapter 3 for more discussion). Managers need to explain the reasons for the need for convergence, and to use shared values to rally staff behind the process.

Convergence is happening in many countries of the world, and this phenomenon will evolve further with time. Ifra's Northrup described convergence as an "established" industry trend that was "no longer just an experiment or fad." In most cases, he said, the more successful media companies were also "the more converging media companies" (Northrup 2004d). In markets characterized by high competition, convergence may become inevitable as people select their news and information from an increasing variety of sources and as media companies respond to find ways to reach audiences that continue to fragment. Specific cultural and legal structures outlined in the first three chapters make convergence possible, and relatively cheap digital tools discussed in chapters 2, 4 and 6 mean that multi-media newsgathering is becoming easier to do. Training and education remain paramount if convergence is to work. Unique content created by talented people will continue to be one of the main differences between high- and mediocre-quality journalism as we enter the knowledge age, especially for organizations seeking to reach the prosperous AB demographic.

Regardless of the form that integration takes, journalists and their managers need to remember that convergence should be about journalism. Both need to ensure that convergence remains a journalistic process led by journalists. The key role of journalism should remain paramount: to weave and maintain the fabric of society and keep that society informed so its citizens can make enlightened decisions.

Glossary

actuality Journalism term that refers to footage or tape of actual events, usually quotes from interview.
AdsML An international standard for digital exchange of advertising information. (See www.adsml.org.)
affiliate TV station not owned by a network that sells a network use of specific time periods for network programs and advertising. Remainder of broadcast day is programmed locally.
American Federation of Television and Radio Artists (AFTRA) Union of broadcasting workers with headquarters in New York.
American Women in Radio and Television (AWRT) An association in Washington, D.C., of women who work in all areas of broadcasting.
anchor Person who coordinates, or anchors, news reports.
assignment desk or news desk Staff responsible for sending camera crews and reporters to cover news events.
assignment editor Person responsible for judging appropriateness of story ideas assigned to reporter for coverage.
banner Print media term for a headline for important story that stretches the width of the page.
broadband services High-speed cable Internet, digital cable, and digital phone services all through a single pipeline.
byline Name of writer or reporter crediting the author, usually printed at the start of story.
circulation Distribution of newspapers, magazines, and other print publications.
clip or clipping Story cut from a publication for storage, or segment cut from a video or audiotape.
close-up A camera technique in which a face fills the frame.
contributing reporter or writer Term often used to describe a freelance writer.
co-op advertising TV advertising paid for jointly by a manufacturer and retailer.

copy editor Last professional to see and approve written material before it is delivered to an audience by a media outlet. Responsible for its legals, accuracy, grammar, and length.

cost per thousand (CPM) Cost of reaching 1,000 homes or individuals with a specific advertising message. CPM is a standard advertising measure to compare the relative cost efficiency of different programs, stations, or media.

cost-per-rating point The cost of reaching one percent of the target audience within a specified geographic area. Used by most media planners in developing and allocating market budgets and setting rating point goals.

cutline Caption to a picture or other graphic element of a story.

cutting Editing tape or film.

dateline A line at the beginning of a printed news story or news release giving the place and date of the story's origin.

dayparts Time segments that divide a television day for advertising scheduling. These segments generally reflect a television station's programming patterns. Comparison of audience estimates between dayparts may indicate differences in size and composition of an available audience.

dead air Technical error in which no signal is fed for broadcast.

demographics Audience composition based on various socio-economic characteristics such as gender, age, income, education, household size, and occupation.

direct broadcast satellite (DBS) A television technology that delivers signals directly from a satellite to a home through the use of a small dish (usually about 18 inches).

dub Most commonly, a copy of a film or tape. It can also mean making a completely new soundtrack, as in dubbing English for a foreign film. A dubbed tape is also called a dupe.

editing In print, the act of rewriting and sometimes cutting print publications; in television, the act of assembling the order of pictures.

editor The person who edits stories (print) or videotape (television).

editorial calendar A list of specific times a publication will report on special sections or news.

editorial Statement of opinion from an editor or publisher found on the op-ed page. A general term for media coverage generated by news staff.

exclusive A news item or feature article that only one newspaper, magazine, or television station carries.

feature A long, probing article or story (as opposed to a news item). Magazines and newspapers have a features department or desk.

Federal Communications Commission (FCC) A US government agency, directly responsible to Congress. The FCC was established by the Communications Act of 1934 and is charged with regulating interstate and international communications by radio, television, wire, satellite, and cable. The FCC's jurisdiction covers the 50 states, the District of Columbia, and US possessions.

feed To send a program or signal. For example, feeding a program from one station via satellite to another station.

first generation Original film or tape. Dubs or dupes are made from it. Applies to analog tapes.

footage A selection or sections of film already shot.

freelancer A writer who sells writing services and is not under regular contract to any one publication.

frequency In print, the number of times a publication comes out in a given period of time, such as daily, weekly, or monthly. In television, the average number of times an accumulated audience has the opportunity to be exposed to advertisements, a particular program, or program schedule, within a measured period of time.

gatekeeping Selection or rejection of news. An editor decides what is going to be covered; an organizer briefs the camera crews; the reporter decides what "angle" the story will be covered from. These people are all gatekeepers.

ghostwriter A person who writes articles or speeches for another person who claims authorship.

green room A small room near a broadcast studio in which program guests wait before they are interviewed.

high definition television (HDTV) Technical systems that provide a finer and wider TV picture and usually twice as many scanning lines as standard TV. Provides the best quality picture and sound, plus the chance to deliver large amounts of data.

hot Refers to a microphone that is switched on.

households using television (HUT) The percentage of all television households in a survey area with one or more sets in use during a specific time period.

in the can A show or portion of it that is complete and ready to broadcast. Before digital delivery, the tape or film was literally stored in a can.

independent station Station not affiliated with any network. Usually refers to commercial stations.

infomercial Television commercial similar in appearance to a news program or talk show format, usually 30 minutes in length.
lapel or lavaliere mike Small microphone attached to a lapel or blouse.
lead Opening sentence or paragraph of a news story.
lead time Time that reporters and producers need to prepare stories and information for publication or broadcast.
leader or leader blank Blank film or tape used at the "head" or start of a program. Can be used to thread a projector or tape machine so the show can go on the air as soon as the machine starts running.
lead-in Program that immediately precedes another program on the same station or network.
letters to the editor Opportunity for the public to discuss or criticize an article (print).
local spot The advertising purchased in a market and aimed only at the audience in that market.
long shot Video and film term for long-distance filming; good for location.
masthead List of editors, publishers, and senior reporters in each publication's issue. It includes an address and telephone number
medium close-up Tight camera angle that frames the subject's head and shoulders
medium shot Camera angle that frames a whole person with some activity visible.
medium wide shot Wider camera angle framing some activity.
mobile unit Television equipment used outside the studio, as at a football game.
mult, mult box or multiplexer Device connected to the main microphone at a news event, which individual broadcast journalists or crews can plug or "patch" into, eliminating the need for a forest of mikes at the podium. Each unit usually handles 12 to 24 separate lines.
National Association of Broadcasters (NAB) A non-profit organization supported by TV and radio broadcasters that lobbies legislative bodies and other interested parties to show that broadcasters can maintain adequate standards without government interference.
network Chain of broadcast or radio stations controlled and operated as a unit, often using the same editorial material.
news Breaking news, immediate media coverage as events occur.
news desk Newspaper desk that assigns reporters to stories.
news feature Special story or article in a print publication or broadcast program that goes into more detail than a regular news item.

news management A term used to describe the way that individuals or organizations attempt to control the flow of news to the media and to "set the agenda" for the media. This might involve issuing a press release which is embargoed, holding press conferences timed to catch the lunch-time and early-evening news, or staging an event that is big or unusual enough to grab media attention.

news peg or hook Aspect of a story that makes it newsworthy, important, or interesting. The news peg is often based on news values such as timeliness, wide impact, or effect on community.

news values Criteria journalists and editors use to determine whether an event is "newsworthy."

Nielsen Media Research (NMR) Firm involved in local and national measurement of TV audiences; also involved in other research activities.

Nielsen Station Index (NSI) Reports measuring local market audiences.

Nielsen Television Index (NTI) National network audience measurement reports.

op-ed Article written by an expert that is positioned on the page opposite the editorial page. Not to be confused with letters to the editor.

open-mike A microphone in use.

package In television news, a collection of interviews, voice-overs, and video footage assembled to tell a story.

panning Moving the camera from left to right or right to left.

ParkerVision Remote controlled cameras operated from a studio control room.

pay-per-view TV (PPV) System in which payment is made for a single showing of a program. Subscribers of the pay-television company can phone in "orders" prior to a showing, activate the system—that is, clear the scrambled channel—or press a button to utilize two-way equipment that activates the system.

penetration Proportion of households owning televisions or subscribing to cable.

periodical Publication circulated at regular intervals, such as a weekly or monthly.

personal people meter (PPM) Pager-sized device worn by a consumer throughout the day. Automatically detects inaudible codes that radio and television broadcasters and cable networks embed in the audio portion of their programming. Used to monitor individual media usage.

personal video recorder (PVR) Digital devices that use a hard drive instead of videotape as the recording medium. PVRs blend the features of a VCR with the programmability and storage of a computer. Best known version is TiVo.

piggyback Back-to-back scheduling of two or more brand commercials of one advertiser in network or spot positions.

pool Camera crew and reporter(s) assigned to cover a story or event on behalf of all media and to share materials. Often used when tight space and security are considerations.

psychographics Audience analysis on the basis of psychological factors such as lifestyles, values, and interests and how they affect purchase behavior.

producer Person in charge of coordinating all details pertaining to a television or radio program. Also used to describe an online editor.

public service broadcasting Process in which the media's role is to inform and educate as well as entertain the public. This stems from the notion that broadcasting's function is not simply to satisfy commercial interests by giving the public what they want to maximize audience figures.

rating Percentage of total households or population owning televisions who are tuned to a particular program or station at a specific time. For example, a nine rating for women 18–49 means 9 percent of all women aged 18–49 in the defined geographic area were viewing that station or program.

rating point Value equal to one percent of a population.

reach Number of unduplicated households or people exposed to a program, group of programs, or an advertiser's schedule over a specific time period; geographic area of the audience and the number of readers, listeners, or viewers who can access the media in any region. Reach and frequency formulas: Reach x Frequency = Gross Rating Points.

remote Broadcast coming from outside the studio.

reverse Camera shot taken from a reverse angle, for example, from over the shoulder or behind the interview subject, with the reporter in the frame. Used to link different parts of an interview. Often photojournalists take "cover" shots of an audience at a press conference to give editors cut-aways.

round-up story Story geared to look back at what has happened over a specific period of time, such as the previous year or quarter.

running time Duration of a show from start to finish; often abbreviated TRT (total running time).

satellite dish (C-Band/KU Band), "big dish" Equipment that allows a household to receive transmissions via satellite by a 1-meter to 3-meter dish (one yard or three yards).

Glossary

satellite master antenna (SMATV) Form of equipment that serves housing complexes such as apartment blocks and hotels; the signal is received via satellite and distributed by coaxial cable.
satellite station Station that has agreed to rebroadcast the transmission of another station (generally operating in a larger nearby market) to an area that cannot otherwise be served by that station.
satellite tour A feed from one point of origination to various downlink sites.
sequence Shooting and editing that unfolds action in such a way that it makes sense to a viewer, and also compresses time.
share Audience during the average minute of a program, expressed as a percentage of households using TV; computed by dividing average audience rating by households using TV.
sidebar Short piece of text accompanying a larger story, often with a human interest angle; usually blocked off from the main text.
sitting mike A microphone on a table.
sound effects Music, explosions, and the like.
spot news News of immediate interest.
stage manager Individual on the set of a studio who gives cues and provides liaison with the control room. No longer needed with ParkerVision.
Stand-up Commentary or report by a TV correspondent seen on camera, usually at the scene of the action. Used to open, close, or bridge the elements of a report.
sting Piece of music used to alert audiences to a specific program or event.
super Image superimposed over another image. Often the subject of an interview will be "supered" with identifying information; also called "titles" or "lower thirds" because of the position of the words.
superstation Station that provides satellite transmission of its signal to cable systems throughout the country. The extended coverage allows the superstation to claim increased numbers of viewers.
sweeps Four-week periods (known as "sweep months") in which NSI surveys all television markets (generally November, February, May, and July).
syndicated columnist Person hired by publications or broadcast organizations to produce written or spoken commentary about specific feature subjects. A syndicated print column is usually published in a wide variety of newspapers, magazines, or on many local networks.
talent Television journalist's term for an interviewee.
talking head A television shot that shows only the upper shoulders, neck, and face of the person being interviewed. Usually accompanied by a

computer-generated sign that appears midchest identifying the person and his or her company.

target audience Audience most desired by advertisers in terms of potential product or service usage and revenue potential.

television households An estimate of the number of households that have one or more television sets.

tilting Moving the camera up and down.

tracking shot Camera shot that moves with the action.

ultra high frequency (UHF) Area of the broadcast spectrum that carries television signals for stations with channels 14 through 83.

very high frequency (VHF) Area of the broadcast spectrum that carries television signals for stations with channels 2 through 13.

Video new release Television equivalent of a news release.

voice-over (v/o) Reporter's voice played along with visuals to tell a news story.

wallpaper Narration leading into or out of a video clip. Narration over visuals, written by anyone, read by anchor.

wire service News stories and features sent by direct line to subscriber or member newspapers and radio and television stations. The Associated Press is the world's biggest.

zoom in Camera action in which the lens optically enlarges the image, focuses attention on detail.

zoom out Camera action in which the lens optically widens the image, reveals scene or activity.

References

ACNeilsen (2003). "Up close and personal—Nielsen media research reveals what Singaporeans read, watch and listen." http://www.acnielsen.com.sg/news.asp? newsID=134 [accessed 29 April 2004].

Aeria, M. (2002). Comment made during the third international Ifra seminar, Defining Convergence, 13 November 2002.

Allen, T. (1977). *Managing the Flow of Technology*. Cambridge, Massachusetts: MIT Press.

Altine, K. (2004). Interview at the *San Francisco Chronicle*, 12 April 2004.

Anderson, G. B. (2001). Presentation to the World Editors' Forum in Hong Kong, 5 June 2001.

Anonymous (2003). "Inside newspaper culture," published by the Readership Institute at http://www.readership.org/culture_management/culture/inside_culture.htm [accessed 20 June 2004].

Anonymous (2004). "Jones urges newspaper chiefs to innovate, seek improvements," in *The Greeneville Sun*, 23 April 2004, A1.

API (2003). "Seventy percent of media consumers use multiple forms of media at the same time," published 24 March 2004 at http://www.mediacenter.org/content/3673.cfm [accessed 24 March 2004].

API (2004). "Convergence Tracker," at http://www.americanpressinstitute.org/ convergencetracker [accessed 1 September 2004].

Auletta, K. (2001). *Backstory: Inside the Business of News*. London: Profile Books.

Barnes, A. (2002). "Differences in craft could threaten quality," transcript of Journalism and Business Values conference published 24 January 2002 at http://www.poynter.org/ content/content_view.asp?id=3488 [accessed 22 June 2004].

Beeston, J. (2001). "CNN Hong Kong: The digital news room." Presentation to the World Association of Newspapers in Hong Kong, 5 June 2001.

Bennett, R. (2000). "Horizon watching: Seizing new opportunities," Paper presented to Ifra's "Beyond the Printed Word," conference, Amsterdam, 9–12 October 2000.

Bergman, C. (2004). "MSNBC.com sees surge in video ads," published 16 July 2004 in Lost Remote at http://www.lostremote.com/archives/001676.html [accessed 16 July 2004].

Bierbauer, C. (2004). Interview in Columbia, South Carolina, 15 October 2004.

Bierhoff, J.; Deuze, M.; and de Vreese, C. (2000). "Media innovation, professional debate and media training: A European analysis," published by the European Journalism Center, December 2000.

Brand, S. (1987). *The Media Lab: Inventing the Future at MIT*. New York: Viking.

Broadcast Engineering (2004). "Video journalists extend reach of BBC without adding costs," published 18 February 2004 in News Technology Update at http://newstechnology update.broadcastengineering.com/february_18/index.htm [accessed 23 April 2004].

Brock, G. (2001). Interview at *The Times* in London, 19 April 2001.

Brown, J. (2004). Interview at the *San Francisco Chronicle*, 12 April 2004.

Bughin, J. (2004). *The McKinsey Quarterly*, number 3.

Burt, T. (2004). "Napster and News Corp arm in link-up," published 21 July 2004 in the *Financial Times*, 4.

Burt, T. and London, S. (2004). "Profits in the age of an audience of one," published 16 April 2004 in the *Financial Times*, 6.

Business Times (2004). "Everybody now is a member of the media," published 17 July 2004 at http://business-times.asia1.com.sg/story/0,4567,122779,00.html [accessed 18 July 2004].

11Campbell, C. (2004). "Taking the initiative to blaze the trail ahead," in *newspaper techniques*, April 2004, 8–9.

Carr, F. (2002a). "Common convergence questions," published 2 May 2002 on Poynter Online at http://www.poynter.org/content/content_view.asp?id=9614 [accessed 2 August 2004].

Carr, F. (2002b). "The truth about convergence," published 2 May 2002 on Poynter Online at http://www.poynter.org/content/content_view.asp?id=9677 [accessed 9 July 2004].

Carr, F. (2004). Interview in Tampa, Florida, 7 February 2004.

Chang, D-W. (2001a). "Remaking a newspaper: Maeil Business Newspaper Case." Presentation to the World Association of Newspapers in Hong Kong, 6 June 2001.

Chang, D-W. (2001b). Email interview 4 September 2001.

Cheung, P. (2001). Interview in Hong Kong, 7 June 2001. Also quoted in Loh, P. (2000). "Credibility of content guides *Ming Pao*'s move to multiple media," *newspaper techniques*, December, 46–48.

Cole, D. (2000). "Convergence: Whether it's in video, audio or war-torn East Timor, it's still journalism," Cole Papers, April 2000, volume 11, number 4.

Colon, A. (2000). "The multi-media newsroom: Three organizations aim for convergence in newly designed Tampa headquarters," in *Columbia Journalism Review*, May/June 2000.

Compaine, B. (2002). Personal communication in Turku, Finland, 10 May 2002.

Covington, R. (2004a). "News flow editor focuses on journalism, not delivery method," published 2 June 2004 in *The Convergence Newsletter*, volume 1, number 11 at http://www.jour.sc.edu/news/convergence/index.html [accessed 17 July 2004].

Covington, R. (2004b). "Old media, old ideas; new media, new hope," published 1 September 2004 in *The Convergence Newsletter*, volume 2, number 3 at http://www.jour.sc.edu/ news/convergence/issue14.html [accessed 11 September 2004].

Crosby, V. (2003). Comment as chair of media technology panel, Online News Association annual conference, Chicago, 15 November 2003.

Curley, R. (2003). Presentation to Online News Association, Chicago, 15 November 2003. Interview in Chicago the same day.

Curley, R. (2004a). Interview in Lawrence, Kansas, 26 April 2004; also email interview, 5 June 2004.

Curley, R. (2004b). Email interviews, 5 June 2004 through 15 August 2004.

Dacruz, M. (2004). "Convergence key to media survival, CanWest exec says," published 25 March 2004 in *The Financial Post* at http://www.canada.com/montreal/specials/business/story.html?id=6B031A65-9768-4395-9911-38433738A771 [accessed 30 March 2004].

Dailey, L.; Demo, L.; Spillman, M. (2003). "The convergence continuum: A model for studying collaboration between media newsrooms." Paper submitted to the Newspaper Division of the Association for Education in Journalism and Mass Communication, Kansas City, Missouri, 1 August 2003.

Damewood, A. (2004). "*New York Times* publisher shares his vision for the future of journalism." Report on Sulzberger's February 23 Crain lecture "Journalism at the turn of the century," published 24 February 2004 at http://www.medill.northwestern.edu/inside/2004/sulzberger [accessed 1 March 2004].

Davenport, T. and Prusak, L. (1998). *Working Knowledge.* Boston: Harvard Business School Press.

Davies, J. (2004). "Ifra hosts first U.S. study tour for European publishers," in *newspaper techniques,* June 2004, 9.

de Aquino, R. (2002). "The print European landscape in the context of multimedia." Presentation to Mudia (Multimedia in a Digital Age) in Bruges, Belgium in May 2002. Found at http://mudia.ecdc.info/index.html [accessed 23 December 2004].

de Aquino, R.; Bierhoff, J.; Orchard, T.; and Stone, M. (2002). *The European multi-media news landscape.* Maastricht: Mudia project.

Dobrow, L. (2004). "Broadband growth numbers spur contention," published 6 August 2004 in *Media Post,* at http://www.mediapost.com/PrintFriend.cfm?articleId=263077 [accessed 7 August 2004].

Dotinga, R. (2003). "Convergence gains critical mass," published 12 May 2003 in Editor & Publisher Online at http://www.editorandpublisher.com/eandp/news/article_display.jsp?vnu_content_id=1885814 [accessed 23 May 2004].

Dyson, R. (2003). Comments made during panel discussion about the future at the Online News Association's annual conference, Chicago, 15 November 2003.

Elkin, T. (2004). "The art of the cross-media deal," published in *Media Daily News,* 23 August 2004 at http://www.mediapost.com/dtls_dsp_news.cfm?newsId=265321 [accessed 1 September 2004].

Estes, B. (2003). Presentation to Online News Association annual conference, Chicago, 15 November 2003.

Fahim, S. (2004). Interview in New York 14 May 2004 and via email.

Farhi, P. (2003). "Mega-media: Better or more of the same?" in *The Washington Post,* 3 June 2003, C1.

FCC (2004). "Media ownership policy re-examination," published 1 January 2004 at http://www.fcc.gov/ownership [accessed 23 April 2004].

Feaver, D. (2004). Telephone interview, 22 April 2004.

Fidler, R. (1997). *Mediamorphosis: Understanding new media.* New York: Pine Forge Press.

Filak, V. (2003). "Inter-group bias and convergence education: A theoretical discussion of conflict resolution between print and broadcast journalism students." Presentation to the Expanding Convergence conference, University of South Carolina, 8 November 2003.

Filak, V. (2004). "Cultural convergence: Inter-group bias among journalists and its impact on convergence," *Atlantic Journal of Communication,* volume 12, number 4.

Financial Times (2003). "U.S. media deregulation," published 5 September 2003 at http://news.ft.com/servlet/ContentServer?pagename=FT.com/StoryFT/FullStory&c=StoryFT&cid=1051390386405&p=1012571727088 [accessed 22 June 2004].

Finberg, H. (2002). "Convergence and the Corporate Boardroom." Presentation to the third international Ifra seminar, Defining Convergence, 13 November 2002.

Finberg, H. (2003). "Convergence and changing media corporate culture," on Poynter Online at http://www.poynter.org [accessed 15 June 2004].

Finberg, H. (2004). Comment made during seminar on convergence for journalism educators, Poynter Institute, 10 February 2004.

Fisher, D. (2004). "'Story builder' embodies new roles in evolving newsrooms," published April 2004 in *The Convergence Newsletter*, volume 1, number 9 at http://www.jour.sc.edu/news/ convergence/index.html [accessed 9 July 2004].

Fisher, H. (2004a). "Newsplex Fellow explores convergence," published 2 March 2004 in *The Convergence Newsletter*, volume 1, number 8 at http://www.jour.sc.edu/news/convergence/ index.html [accessed 9 July 2004].

Fisher, H. (2004b). "Newsplex announces 2004 technology picks" published 7 July 2004 in *The Convergence Newsletter*, volume 2, number 1 at http://www.jour.sc.edu/news/ convergence/index.html [accessed 9 July 2004].

Fournier, V. (2002). "Newsplex will be a lab for convergence," in *newspaper techniques*, July/August 2002, 30–31.

Fuller, J. (1996). *News Values: Ideas for an information age*. Chicago: University of Chicago Press.

Fuller, J. (2003). Keynote speech to Online News Association's annual conference, Chicago, 15 November 2003.

Gentry, J. (1999). "The *Orlando Sentinel*. Newspaper of the future: Integrating print, television and Web," in *Making Change*, a report for the American Society of Newspaper Editors, April 1999, 3–9.

Gentry, J. (2000). "Making change in a media environment." Presentation to the American Press Institute in Washington, 8 June 2000.

Gentry, J. (2003). "Convergence case study: LJWorld.com," published 14 April 2003 at http://www. mediacenter.org/content/777.cfm [accessed 15 April 2004].

Gentry, J. (2004a). "Today's convergence newsroom." Presentation to journalism educators at the Poynter Institute, St. Petersburg, Florida, 8 February 2004.

Gentry, J. (2004b). "Toward a cross-platform curriculum." Presentation to journalism educators at the Poynter Institute, St. Petersburg, Florida, 9 February 2004. Also, various email conversations in 2004.

Gentry, J. and Musser, R. (2004). Interview in Lawrence, Kansas, 26 April 2004.

Giner, J. A. (2001a). "From media companies to 'information engines,'" in *Innovations in Newspapers 2001 World Report,* Innovation International, Pamplona, Spain, 28–33. See also http://www.innovacion.com.

Giner, J. A. (2001b). "From newspapers to 24-hour information engines," in *Ideas* magazine, October 2001 at http://www.inma.org [accessed 11 June 2004].

Giner, J. A. (2001c). Comments posted to www.inma.org on 8 March 2001.

Giuliano, V. (2001). Interviews in Boston, 25 March and 15 April 2001.

Glamann, H. (2000). "Equipping employees for change," in *newspaper techniques,* January 2000, 48.

Glaser, M. (2004a). "Lack of unions makes Florida the convergence state," in *Online Journalism Review,* at http://ojr.org/ojr/glaser/1081317274.php [accessed 8 April 2004].

Glaser, M. (2004b). "Business side of convergence has myths, some real benefits," published 19 May 2004 in *Online Journalism Review,* at http://ojr.org/ojr/business/ 1084948706.php [accessed 28 July 2004].

Gordon, R. (2003). "The meanings and implications of convergence," pages 57–73, in *Digital Journalism: Emerging Media and the Changing Horizons of Journalism* edited by Kevin Kawamoto. New York: Rowman & Littlefield Publishers.

Grant, A. (2004a). "Five steps to a convergent newslab," in *The Convergence Newsletter,* volume 2, number 1, published 7 July 2004 at http://www.jour.sc.edu/news/convergence/index.html [accessed 9 July 2004].

Grant, A. (2004b). "Multi-skilled journalists are prepared to tell stories in many forms," published 7 July 2004 in *The Convergence Newsletter,* volume 2, number 1 at http://www.jour.sc.edu/news/convergence/index.html [accessed 9 July 2004].

Haagerup, U. (2002a). Comment during a defining convergence roundtable at the third Ifra international newsroom summit, 13 November 2002.

Haagerup, U. (2002b). Presentation to Ifra's Defining Convergence conference, Columbia, South Carolina, 14 November 2002.

Haagerup, U. (2004). Email interview, 2 February 2004.
Haile, J. (2003). "Creating a change culture: A case study of the *Orlando Sentinel*'s transition to a multi-media company," Inside Out Media. Consultant's report.
Haile, J. (2004). Email interview, 1 September 2004.
Haiman, R. (2001). "Can convergence float?" published 28 February 2001 at Poynter Online at http://www.poynter.org/content/content_view.asp?id=14540 [accessed 14 February 2004]. Repeated during presentation to journalism educators, 10 February 2004.
Han, C. H. (2004). Interview in Muncie, Indiana, 23 April 2004.
Hanafin, T. (2003). Presentation to Online News Association annual conference, Chicago, 15 November 2003.
Harper, P. (2004). "Managing the data deluge: How to stay afloat in life's sea of information," published at http://mobilemomentum.msn.com/article.aspx?aid=1 [accessed 12 June 2004].
Hartigan, J. (2002). "Hartigan hits out at 'dictatorial' newsagents," in PANPA *Bulletin*, September 2002, 5–8.
Hilmer, F. (2004). Interview in Sydney, 8 July 2004.
Hinojosa, M. (2004). Interview in Chicago, 4 April 2004.
Horrocks, P. (2002). Comment during a defining convergence roundtable at the third Ifra international newsroom summit, 13 November 2002.
Huang, E.; Rademakers, L.; Moshood, A. F.; and Dunlap, L. (2004). "Uncovering the quality of converged journalism: A case study of *Tampa Tribune* news stories." Presentation to the annual convention of the Association for Education in Journalism and Mass Communication, Toronto, Canada, 2 August 2004.
Jackson, S. (2001). "Convergent views" in the Media section of *The Australian*, 28 June 2001, 12.
Janischewski, C. (2003). "Strengthening brands through the use of convergence," in *newspaper techniques*, February 2003, 34–35.
Kamaras, C. (2002). "Newspapers rethinking the web," in *newspaper techniques*, July/August 2002, 40–42.
Kaye, K. (2004). "WSJ Online study reveals Internet revolution has reached the corner office," *The Wall Street Journal*, 24 May 2004, C1.
Keichel, W. (2001). "Co-opetition true and false: An alternative to convergence?" Presentation to the World Association of Newspapers in Hong Kong, 6 June 2001.
Khalaf, R. (2003). "Arab TV gears up for fight on home front," in *Financial Times* 2 February 2003, 12.

Khan, M. (2004). "Change is accelerating in ad business, net.marketing speaker says," published 25 February 2004 in *New York Daily News* online [accessed 30 March 2004].

Kieve, J. (1973). *Electric Telegraph: A Social and Economic History.* Devon, UK: David & Charles.

Kirk, J. (2004). "P&G launches major change in media spending," published 15 July 2004 at http://www.chicagotribune.com/business/chi-0407150347jul15,1,325081.story [accessed 20 July 2004].

Kiss, J. (2003a). "Hacks on the beach" published 24 July 2003 in dot-journalism at http://www.journalism.co.uk/news/story691.html [accessed 20 May 2004].

Kiss, J. (2003b). "Wireless fidelity—all the way" published 24 September 2003 in dot-journalism at http://www.journalism.co.uk/news/story725.html [accessed 20 May 2004].

Kiss, J. (2003c). "New software enables laptop broadcasts," published 3 December 2003 on dot-journalism at http://www.journalism.co.uk/news/story775.shtml [accessed 20 May 2004].

Kiss, J. (2004a). "TV team files story wirelessly" published 11 February 2004 in dot-journalism at http://www.journalism.co.uk/news/story814.shtml [accessed 20 May 2004].

Kiss, J. (2004b). "News filed on the spot—thumbs permitting," published 6 May 2004 in dot-journalism at http://www.journalism.co.uk/news/story875.shtml [accessed 20 May 2004].

Knight, K. (2004). Interview at the *Tampa Tribune,* 9 March 2004.

Kolodzy, J. (2003a). "Everything that rises: Media convergence is an opportunity, not a curse," in *Columbia Journalism Review,* volume 42 July/August 2003, 61.

Kolodzy, J. (2003b). Comments made during a panel entitled "What's working in teaching convergent media," at the University of South Carolina at Columbia, 8 November 2003.

Kovach, W. and Rosenstiel, T. (2001). *The Elements of Journalism.* New York: Three Rivers Press.

Landberg, M. (2004). "Oh no! Live TV on cell phones," in *San Jose Mercury News,* http://www.mercurynews.com/mld/mercurynews/business/technology/personal_technology/8019184.htm [accessed 25 February 2004].

Landphair, T. (2004). "Report offers sobering critique of US media," published by Voice of America online at http://www.voanews.com/article.cfm?objectID=AEDFB36F-3A3B-4DEA-BD3938470FEAE0DD [accessed 15 April 2004].

Larsen, J. (2004). Email interview 18 June 2004. Interview in Aarhus, Denmark, 25 November 2004.

Larsen, P. T. (2004). "Multi-media? That's just the consumers," published 1 August 2004 in *Financial Times* online, at http://news.ft.com/cms/s/739b1c9a-e3fe-11d8–9f08–00000e2511c8.html [accessed 2 August 2004].
Lie, R. (2000). Quoted in *The Ifra Trend Report*, number 64, 11 October 2000, 1.
Lim, V. (2004). Interview at WFLA-TV, Tampa, 8 March 2004.
Livingston, K. L. (1996). *The Wired Nation Continent: The Communication Revolution and Federating Australia*. Melbourne: Oxford University Press.
LoCicero, G. (2004a). "News resourcer is key information chief," published 5 May 2004 in *The Convergence Newsletter*, volume 1, number 10 at http://www.jour.sc.edu/news/convergence/index.html [accessed 9 July 2004].
LoCicero, G. (2004b). "Blogs offer journalists a chance to try convergence" published 1 September 2004 in *The Convergence Newsletter*, volume 2, number 3 at http://www.jour.sc.edu/news/convergence/issue14.html [accessed 11 September 2004].
Locin, M. (2003). Interview in Muncie, Indiana, 28 October 2003.
Loh, P. (2000). "Credibility of content guides *Ming Pao*'s move to multiple media," *newspaper techniques*, December, 46–48.
Mandese, J. (2004). "Study: Media overload on the rise," in *Television Week* online. http://www.tvweek.com/planning/051704study.html [accessed 17 May 2004].
Marshall, A. (2004). Email interviews, September 2004.
Maxwell, S. (2004). Interview at the *Orlando Sentinel*, 10 March 2004.
McCartney, R. (2004). "Continuous news at *The Washington Post*." Presentation to the Online News Association annual conference, Los Angeles, 12 November 2004.
McCombs, R. (2004). Phone interview 3 December 2004.
McDonough, S. (2004). "Newspaper association eyes new technology," http://news.yahoo.com/news/story=20040421 [accessed 23 April 2004].
McFarlin, D. (2002). "Convergence breeds stronger journalism," transcript of Journalism and Business Values conference published 24 January 2002 at http://www.poynter.org/content/content_view.asp?id=3488 [accessed 22 June 2004].
McNair, B. (1998). *The Sociology of Journalism*. London: Arnold.
Miller, S. (2004). Interview at *The New York Times*, 10 May 2004.
MUDIA (2002). See http://mudia.ecdc.info/index.html [accessed 1 September 2004].

Myles, P. (2004). Email interviews 1–3 September 2004.
Nachison, A. (2001a). "Good business or good journalism? Lessons from the bleeding edge." Presentation to the World Editors' Forum, Hong Kong, 5 June 2001. Email communication 12 June 2001.
Nachison, A. (2001b). Email interview, 12 June 2001.
Nachison, A. (2002). Comment made during the Ifra seminar, Defining Convergence, 13 November 2002.
Nachison, A. (2004). Telephone interview, 17 February 2004.
Newsplex (2004). "Ifra awards NewsGear 2004 designations," http://www.newsplex.org/ [accessed 29 April 2004].
Newton, E. (2003). "Great demand, weak supply," in the *Learning Newsroom*, published by the American Society of Newspaper Editors, 9.
Nichols, J. (2004). "Newsplex news," published April 2004 in *The Convergence Newsletter*, volume 1, number 9 at http://www.jour.sc.edu/news/convergence/index.html [accessed 9 July 2004].
Nielsen (2004). See http://www.nielsen-netratings.com [accessed 28 March 2004].
Northrup, K. (1996). "The digital newsroom of the future: A new way of managing the news," in *The Seybold Report on Publishing Systems*, volume 25, number 18, 1996, 3–8.
Northrup, K. (1999). Presentation to the Newsroom for a Digital Age conference, 7–8 December 1999, Darmstadt, Germany.
Northrup, K. (2000a). Interview in Melbourne, 23 July 2000.
Northrup, K. (2000b). "New skills needed for today's 'multiple media' stories," in *Bulletin* of the Pacific Area Newspaper Publishers' Association, November 2000, 32–33.
Northrup, K. (2002a). "Building the newsroom of the future—today," in *newspaper techniques*, July/August 2002, 15.
Northrup, K. (2002b). "Much more than just technology," in *newspaper techniques*, July/August 2002, 16–17.
Northrup, K. (2002c). Opening presentation to the Ifra seminar, Defining Convergence, 13 November 2002.
Northrup, K. (2003a). "A growth market," in *newspaper techniques*, October 2003, 3.
Northrup, K. (2003b). Keynote speech at conference to mark first year of Newsplex, Columbia, 7 November 2003.
Northrup, K. (2004a). "Newsplex models mobile publishing" in *The Seybold Report*, 12 April 2004, 16–18.
Northrup, K. (2004b). Email 1 September 2004.

Northrup, K. (2004c). Statement at Newsplex training, 9 September 2004.

Northrup, K. (2004d). *Newsplex Convergence Guides*, 9 September 2004.

O'Loughlin, T. (2004). "Labor will keep the media laws," in *Australian Financial Review*, 7 July 2004, 5.

ONA (2003). Online News Association's annual conference, Chicago, 14–15 November 2004. Also reported by Finberg, H. (2003) in Convergence Chaser, published 16 November 2004 on Poynter Online at http://www.poynter.org [accessed 21 November 2003].

Outing, S. (2000). "The many possible directions of future media," published 6 December 2000 in *Editor & Publisher* online.

Outing, S. (2004a). "Nathaniel's global multi-media road show," published 1 June 2004 in E-Media Tidbits at http://www.poynter.org/profile/profile.asp?user=1648 [accessed 8 August 2004].

Outing, S. (2004b). "The media world flocks to Kansas?" published 30 June 2004 at http://www.poynter.org/column.asp?id=31&aid=67823 [accessed 1 July 2004].

Outing, S. (2004c). "Staff journalists blog, too," published 26 July 2004 at Poynter Online http://www.poynter.org [accessed 28 July 2004].

Papper, R.; Holmes, M; Popovich, M. (2004). "Middletown Media Studies: Media multi-tasking and how much people really use the media," in the *International Digital Media & Arts Association Journal*, volume 1, number 1, 3–55.

Pascual, M. (2003a). "The path towards convergence becoming more clear," in *newspaper techniques*, July/August 2003, 34–35.

Pascual, M. (2003b). "Changing attitudes essential as a first step towards convergence," in *newspaper techniques*, November 2003, 48–49.

Pascual, M. (2004). "'Dayparting' and 'prime-time': two trumps of online advertising," in *newspaper techniques*, April 2004, 20–21.

Patsuris, P. (2004). "The BBC deploys video cell phones," published 5 February 2004 in Forbes Online at http://www.forbes.com/technology/2004/02/05/cx_pp_ii_0204 cameraphone.html [accessed 9 June 2004].

Peers, M. (2004). "Buddy, can you spare some time?" in *The Wall Street Journal*, 26 January 2004, D1.

Penhune, J. (2004). "8.5 million US homes to add broadband in 2004," press release found at http://www.strategyanalytics.com/press/PR00118.htm [accessed 18 May 2004].

Pew (2004). "Home broadband adoption has increased 60% in past year and use of DSL lines is surging," at http://www.pewinternet.org/pdfs/PIP_Broadband04.DataMemo.pdf [accessed 30 April 2004].

Pfeiffer, A. (2000). "Emerging opportunities in the expanding media space," found at www.ifra.com/NewsFeed.nsf [accessed 20 February 2001].

Phillips, M. (2003). "Embedded reporters tell campaign tales: Iraq war concept of 'living' with the story lands MSNBC's Schein a road trip with Dean," published 29 December 2003 at Wall Street Journal Online http://wsj.com/article/0,,SB107265607546436600,00.html?mod=mm%5Fhs%5Fmedia [accessed 10 January 2004].

Pitts, G. (2002). *Kings of Convergence: The fight for control of Canada's media*. Random House.

Pownall, E. (1973). *The Singing Wire*. Sydney: Collins.

Project for Excellence in Journalism (2004). "The state of the news media 2004," released April 2004 and published at http://www.state-ofthenewsmedia.org/ [accessed 17 May 2004].

Pryor, L. (2004). Email interview 2 August 2004.

Quicklink (2004). Information on News Broadcaster available at http://194.162.230.14/news_broadcaster.html [accessed 12 June 2004].

Quinn, S. (2002). *Knowledge Management in the Digital Newsroom*. Oxford: Focal Press.

Quinn, S. (2004). Observer at Newsplex workshop in Columbia, South Carolina, 7–9 September 2004.

Rainie, L. (2005). "Blog readership increases 58% in 2004" published in Pew Internet & American Life Project, January.

Reddick, R. and King, E. (1995). *The Online Journalist*. Fort Worth: Harcourt Brace & Company.

Redmont, D. (2004). "Newspapers see danger in text messaging," published 8 May 2004 at http://www.eweek.com/article2/0,1759,1588646,00.asp [accessed 19 May 2004].

Reuters (2004). "Web access at 75 percent," *Wired* magazine. http://www.wired.com/news/culture/0,1284,62712,00.html [accessed 15 April 2004].

Rogers, E. (1986). *Communication Technology: The new media in society*. New York: The Free Press.

Rogers, E. (1995). *Diffusion of Innovations*, fourth edition. New York: The Free Press.

Romaner, M. (2002). Comment made during the Ifra seminar, Defining Convergence, 13 November 2002.

Roper, D. (2002a). "Convergence extends to small market paper," in *newspaper techniques*, March 2002, 38–40.

Roper, D. (2002b). "Directorate members plan to take full advantage of proving ground," in *newspaper techniques*, July/August 2002, 22–28.
Roper, D. (2003a). "Communication helps cure convergence change," in *newspaper techniques*, February 2003, 30–32.
Roper, D. (2003b). "Kansas paper proves size doesn't matter," in *newspaper techniques*, December 2003, 14–17.
Rosenbush, S. (2004). "Broadband: What's the holdup?" published 1 March 2004 in *Business Week* at http://www.businessweek.com/magazine/content/04_09/b3872049.htm [accessed 15 April 2004].
Russin, J. (2003). Comment during convergence panel discussion at Online News Association annual conference, Chicago, 15 November 2003.
Saffo, P. (1992). "Paul Saffo and the 30-year rule," in *Design World*, volume 24. Also, interview in Dubai, 20 December 2002.
Sambrook, R. (2004). Speech by BBC head of news Richard Sambrook at a BBC conference on "Future Thinking," in London, 25 March 2004.
Sampson, A. (1996) *Company man: The rise and fall of corporate life*. New York: Random House.
Sandeen, R. (2000). "A look at media convergence: How much multimedia should students learn?" published 27 March 2000 in American Society of Newspaper Editors website at http://www.asne.org/kiosk/editor/00.march/sandeen1.htm [accessed 25 June 2004].
Schroeder, C. (2004). "Is this the future of journalism?" published 18 June 2004 in *Newsweek* at http://www.msnbc.msn.com/id/5240584/site/newsweek [accessed 22 June 2004].
Seely Brown, J. and Duguid, P. (2000). *The Social Life of Information*. Boston: Harvard Business School Press.
Sellen, A. and Harper, R. (2002). *The Myth of the Paperless Office*, Cambridge, Massachusetts: MIT Press.
Shearer, E. (2003). "Convergence," published 27 August 2003 by the American Society of Newspaper Editors at http://www.asne.org/index.cfm?id=4941 [accessed 26 March 2004].
Shenk, D. (1997). *Data Smog: Surviving the Information Glut*. San Francisco: Harper Edge.
Shim, R. (2004). "Powell: Wireless vital to broadband future," published in CNET News at http://news.com/2100-1034-5217044.html [accessed 10 May 2004].

Singer, J. (2003a). "Strange bedfellows: The diffusion of convergence in four news organizations." Presentation to the Media Management and Economics Division, annual conference of the Association for Education in Journalism and Mass Communication, Kansas City, 1 August 2003.

Singer, J. (2003b). "The sociology of convergence: Challenges and change in newspaper news work." Paper presented to the Newspaper Division, Association for Education in Journalism and Mass Communication, Kansas City, 1 August 2003.

Skreien, N. (2003). "Two cases of 'information engines' in Denmark and Finland." *Innovations in Newspapers World Report 2003,* World Association of Newspapers, 2003, 40–47.

Standage, T. (1998). *The Victorian Internet: The remarkable story of the telegraph and the nineteenth century's on-line pioneers.* New York: Walker and Co.

Stephens, M. (1989). *A History of News: From the Drum to the Satellite.* New York: Penguin.

Stevens, J. (2002). "Backpack journalism is here to stay," published 4 February 2002 in *Online Journalism Review,* at http://www.ojr.org/ojr/workplace/1017771575.php [accessed 2 September 2004].

Stewart, T. (2000a). "Water the grass, don't mow, and wait for lightening to strike," in *Fortune,* 24 July 2000. Found online at http://www.fortune.com

Stewart, T. (2000b). "The Leading Edge: The house that knowledge built," in *Fortune,* 2 October 2000. Found online at http://www.fortune.com

Stewart, T. (2001a). "Want innovation," in *Fortune,* 5 March 2001. Found online at http://www.fortune.com

Stewart, T. (2001b). "Intellectual capital: Ten years later, how far we've come," in *Fortune,* 28 May 2001. Found online at http://www.fortune.com

Stone, M. (2002a). "The backpack journalist is a 'mush of mediocrity,'" published 4 February 2002 in *Online Journalism Review,* at http://www.ojr.org/ojr/workplace /1017771634.php [accessed 2 September 2004].

Stone, M. (2002b). "Multi-media integration is here to stay," in *Online Newspapers and Multimedia Newsrooms,* number 4, April 2002, 1–2.

Stone, M. (2003a). *Embracing the power of multi-media advertising.* International Newspaper Marketing Association, Dallas.

Stone, M. (2003b). "Cross-media selling packs a punch," in *newspaper techniques*, October 2003, 32–33. http://www.newsandtech.com/issues/2004/03-04/ifra/03-04_crossmedia.htm.

Stone, M. (2004). "Approaches vary, but newsrooms finding benefits of convergence," in *newspaper techniques*, March 2004, 36–39.

Sveiby, K. E. (1996). "Transfer of knowledge and the information processing professions," in *European Management Journal*, volume 1–14, number 4, August 1996, 379–88.

The New York Times. (2004). "Overnight a done deal is undone," published 18 February 2004 in *The New York Times*, A1. See http://www.nytimes.com/imagepages/2004/02/17/business/18WIRE.chartA1.jpg.html [accessed 2 August 2004].

Thelen, G. (2000a). Presentation to American Press Institute workshop for journalism educators, 8 June 2000.

Thelen, G. (2000b). Interview in Washington, 8 June 2000.

Thelen, G. (2000c). "Tampa's convergence lessons," published 1 July 2000 at Poynter Online at http://www.poynter.org [accessed 12 March 2004].

Thelen, G. (2002a). "A renewed Tribune to serve you better," at *Tampa Tribune* online http://promos.tampatrib.com/promos/readersguide/gil.htm [accessed 24 August 2002].

Thelen, G. (2002b). "Convergence is coming," *Quill*, volume 32, July/August, 16.

Thelen, G. (2002c). Comment made during the Ifra seminar, Defining Convergence, 13 November 2002.

Thelen, G. (2004a). Keynote speech to journalism educators at southeast colloquium of Association for Educators in Journalism and Mass Communication, Tampa, Florida, 5 March 2004.

Thelen, G. (2004b). Interview in Tampa, 5 March 2004.

Thurow, L. (1999). *Building Wealth: The new rules for individuals, companies and nations in a knowledge-based economy*. New York: HarperCollins.

Tiffen, R. and Gittins, R. (2004). *How Australia Compares*. Melbourne: Cambridge University Press.

Tompkins, A. (2002). "Q&A with Mark Stencel," published 30 April 2002 on Poynter Online at http://www.poynter.org/content/content_view.asp?id=3632 [accessed 2 July 2004].

Tyner, H. (2002). Comment made during the Ifra seminar, Defining Convergence, 13 November 2002.

Tyner, H. (2004). Telephone and email interviews, October and November 2004.

Valjakka, A. (2001). "Cost-effective multi-media by *Turun Sanomat.*" Presentation to the World Editors' Forum, Hong Kong, 5 June 2001.

Valjakka, A. (2002a). "Convergence by the numbers." Presentation at the Ifra seminar, Defining Convergence, 13 November 2002.

Valjakka, A. (2002b). Interview in Dubai, 23 December 2002.

Valjakka, A. (2004a). Email interview, 1 February 2004.

Valjakka, A. (2004b). Email interview, 4 September 2004.

Vanjoki, A. (2003). "Sell newspapers to kids on their cell phones," http://www.centerformediaresearch.com/cfmr_brief.cfm?fnl=0310 22 [accessed 22 October 2003].

Veseling, B. (2000). "Flexibility the key in the multi-media world," in PANPA *Bulletin,* November 2000, 20–23.

Veseling, B. (2003). "*Manchester Evening News* builds on Newsplex convergence training," in *newspaper techniques,* September 2003, 102–3.

Wells, J. (2003). "What's in a name," *Toronto Star,* 5 April 2003, A1.

Wheeler, K. (2004a). Interview in St. Petersburg, Florida, 10 February 2004.

Wheeler, K. (2004b). Interview at the *Orlando Sentinel,* 10 March 2004.

White, D. (2001). "How to focus, interview, and get the story told," published 19 October 2001 on Poynter Online at http://www.poynter.org/content/content_view.asp?id=4769 [accessed 12 April 2004].

Wolpe, B. (2002). "Submission by John Fairfax Holdings Limited on the Broadcasting Services Amendment (Media Ownership) Bill 2002," 19 April 2002. Available on the Fairfax corporate website http://www.fxj.com.au [accessed 16 July 2004].

Index

A

Aalto, Alvar, 123
Ability to respond, 202
Adplexing, 78
Advertising Age, 81
Advertising and convergence, 17
Aeria, Michael, 5
Aftenposten, 22, 127
Aftonbladet, 55, 127
Age, The, 53, 67
Altine, Kenn, 56
American Press Institute, 19, 198
Anderson, Gloria Brown, 45, 208
AOL-Time Warner, 7
Arizona Republic, 78
Assignment editor and convergence, 86
Assignment editor, role of, 86
Audience fragmentation, 30
Audience, importance of, 151
Auletta, Ken, 118
Australian Broadcasting Corporation, 68
Australian Financial Review, The, 67
Australian, The, 67

B

Ball State University, 12
Barnes, Andy, 65
Baumgart, Hans-Dieter, 66
BBC Interactive, 37, 194–5,
BBC, 35, 36, 70, 130–1, 138, 192, 199
Beeston, John, 11–2, 145, 151
Belo Corporation, 46, 78, 79
Bierbauer, Charles, 4, 72, 87
Bigman, Dan, 120
Blackberries for newsgathering, 218
Blogs and convergence, 110
Blogs, 110, 114–5, 197
Boat-car, 11
Boston.com, 83
Bottcher, Gunther, 72
Brand, Stewart, 7
Broadband and convergence, 33–5
Broadband penetration, 183–4
Broadband wireless, 34
Broadcast markets, 220
Brock, George, 70
Brown, Joe, 16–7, 56
Budget Bank (software), 23, 54
Burr, John, 42
Business models and convergence, 16–18, 68–9, 205
Business models, 16–8
Business opportunities and convergence, 65

C

Camilleri, Rick, 2, 75, 77
Campaign Embed, 140
CanWest Global Communications, 2
Carr, Forrest, 23, 51, 119
Case, Steve, 7
Cassar, Kenneth, 33
Cast, Carter, 81
CCI-Europe, 23, 54, 218
Cell phones for reaching audiences, 190
Cell-phone photo-journalism, 35, 99, 124, 143
Cell-phone reporting, 92–3, 140, 141–4, 189
Central Florida News 13, 18
Chang, Dae-Whan, 128–9, 215
Change agents, 168–70
Channel M TV, 130
Cheung, Paul, 22, 46, 49, 58, 127–8, 132, 202, 207, 213, 215
Chicago Tribune, The, 20, 28, 29, 117, 196, 213–4
Chief knowledge officers, 169
Chinni, Dante, 31
Choate, Mark, 62
Cingular-AT&T merger, 35
Citizen journalists, 185
Citygate, 73, 188
Coats, Rusty, 82
Colon, Aly, 119
Communication, need for, 15, 202, 206–9
Communications Industry Forecast, 20–21
Communities and convergence, 111–3, 114–5
Communities of interest, 166–7
Compaine, Ben, 191
Continuous news desk, 120
Control Tower (software), 218
Convenience, significance of, 27–28
Convergence "must enter" business, 126
Convergence and "eyeballs", 20, 186
Convergence and advertising, 17
Convergence and assignment editor, 86
Convergence and blogs, 110
Convergence and broadband, 33–5
Convergence and business models, 16–18, 68–9, 205
Convergence and business opportunities, 65
Convergence and common ownership, 50
Convergence and common ownership, 50
Convergence and communities, 111–3, 114–5
Convergence and cooperation, 128
Convergence and cost cutting, 65–6
Convergence and cost-effective newsgathering, 125
Convergence and cultural factors, 23, 47, 48, 56–7
Convergence and databases, 15, 54–5, 115–6, 174, 218–9
Convergence and digital tools, 18–9, 124
Convergence and dissenting voices, 219–20
Convergence and FCC, 39–40
Convergence and fear, 220

Convergence and legal structures, 38–41
Convergence and local content, 73, 111–3, 121–2
Convergence and mindset, 6, 15, 22, 63–6, 86–7, 131–2, 154, 157, 158–9, 162, 193, 205, 206
Convergence and multi-media advertising, 70, 77–83
Convergence and new media, 76
Convergence and new reporting roles, 91–7
Convergence and organizational strategy, 44–45
Convergence and priorities, 202
Convergence and promotion, 65
Convergence and proximity, 57, 159
Convergence and quality content, 66, 113–4, 172, 175, 194, 196, 220–21
Convergence and reporting, 86
Convergence and role of managers, 24–5
Convergence and shared values, 52–3, 206
Convergence and sport, 112–3
Convergence and technology, 32–5, 144–5, 162–3
Convergence and training, 108, 163, 198–9
Convergence and trust, 167–8
Convergence and values, 202, 211
Convergence as change management, 62–3
Convergence as evolution, 50
Convergence as growth strategy, 63–64
Convergence continuum, 12–13
Convergence in Singapore, 38, 187
Convergence Monitor, 132
Convergence not "shovelware", 66, 112
Convergence, "easy" versus "difficult", 42, 116
Convergence, convincing journalists, 123
Convergence, culture differences, 5
Convergence, definitions, 8
Convergence, forces driving it, 20–2
Convergence, how to foster, 22–4
Convergence, Middle East, 3
Convergence, need for definition, 6
Convergence, South America, 3
Convergence, Southeast Asia, 2
Convergence, symbols of, 216
Convergence, variables, 5
Cooperation and convergence, 128
Co-opertition, 13
Cost cutting and convergence, 65–6
Cost-effective newsgathering and convergence, 125
Covington, Randy, 92–3, 141–2, 148
Coyle, Mark, 131
Creative abrasion, 174
Cultural factors and convergence, 23, 47, 48, 56–7
Curley, Rob, 44, 50–1, 54, 57, 103, 106–16, 133, 160–1, 216

D

Dailey, Larry, 12–4
Dallas Morning News, The, 46
Danish Center for Journalism Education, 126

Danish journalists and training, 24
Data smog, 29, 193
Databases and convergence, 15, 54–5, 115–6, 174, 218–9
Databases, 15, 174
de Aquino, Ruth, 21–2, 43, 58, 186–7
Definitions of convergence, 8
Digital tools and convergence, 18–9, 124
Dissenting voices and convergence, 219–20
Dyche, Justin, 35–6
Dyson, Esther, 193

E

"Easy" versus "difficult" convergence, 42, 116
Editorial independence, 45
Electronic paper, 187, 191
Eliot, T.S., 28
Elkin, Tobi, 80
Embracing the Power of Multi-Media Advertising, 78
Ericsson, 188
Estes, Ben, 83, 195–6
Exabytes of information, 28
Extended news desk, 120
Eyeballs and convergence, 20, 186

F

Fahim, Saf, 145–6, 173, 215
FCC and convergence, 39–40
Fear and convergence, 220
Feaver, Doug, 121
Federal Communications Commission, 34, 39–40
Felling, Matthew, 31
Fidler, Roger, 25, 76
Filak, Vince, 210
Final Cut Pro (software), 37, 140
Finberg, Howard, 4, 23–4, 31–2, 47, 48
Fisher, Doug, 93–4
Florida Today, 10
Forces driving convergence, 20–2
Forrester Research, 34–5
Fostering convergence, 22–4
Fuller, Jack, 9, 29, 71, 212
Fulton, Katharine, 19

G

Garcia, Mario, 107
Gardner, Ann, 50, 55, 107, 109, 110, 111
Gentry, James, 4, 16, 41, 45, 74, 107–8, 110, 133, 201, 208–9, 211
Geography and flow of information, 57, 159, 168
Giner, Juan Antonio, 2, 3, 14–5, 31, 49, 72
Giuliano, Vince, 157
Gordon, Rich, 9–12, 75, 119
Gowing, Nik, 180
Granatino, John, 46
Grant, Augie, 96–7
Guardian Media Group, 129
Guardian, The, 139

H

Haagerup, Ulrik, 6, 23, 72, 126, 127
Haile, John, 5, 17–8, 47, 52, 158, 205
Haile-Gentry consultancy, 18
Haiman, Bob, 11, 16
Hanafin, Teresa, 83
Harper, Philipp, 29
Hartigan, John, 162
Harvard Business Review, 13
Hawke-Keating Labor party, Australia, 40–1
Healy, Jane, 50
Helicopter vision (for planning), 203
Henry, Mike, 30
Hilmer, Fred, 66–9, 76, 83, 117
Hinojosa, Mark, 28, 100–1, 118, 214, 215
Hofer, Jochen, 143
Horrocks, Paul, 5, 32, 129–30
Huang, Edgar, 66
Huber, Red, 100

I

Ideas magazine, 2
Impact of innovative technologies, 187–90
InDesign (Adobe software), 148
Information overload, 29, 193
Information-gathering convergence, 10
Innovation theory, 209
Inspector Gadget, 10–1, 14, 219
International Newspaper Marketing Association (INMA), 2, 77
Internet and American homes, 34–5
Internet as photocopier, 156
Internet, dial-up, 35
Internet, executives' use of, 30–1

J

Janischewski, Charlotte, 63
Jonason, Niklas, 73, 74, 188
Journalists' role in society, 221
Jungkvist, Kalle, 127

K

Kamaras, Constantine, 63,
Ketonen, Keijo, 70, 126
Knight, Ken, 53, 54, 57, 134, 160
Knowledge and brand, 157
Knowledge and culture, 170–1
Knowledge creation, 156, 168
Knowledge integrators, 157
Knowledge management, 154, 156–8
Knowledge maps, 171–2
Knowledge sharing, how to promote, 166
Knowledge "talk rooms", 168
Knowledge versus information, 155
Knowledge, explicit versus tacit, 156
Knowledge-based economy, 128–9
Kolodzy, Janet, 47, 69–70, 91
Kovach and Rosenstiel, 25

L

Laptop newsgathering (LNG), 138
Larsen, Jan, 159, 173–4, 175
Latham, Mark, 41
Lavine, John, 67
Learning environments, 159–60
Learning is demand driven, 165–6
Learning newsroom, 198–9
Legal structures and convergence, 38–41
Levitt, Leon, 78
Lie, Rolf, 22, 127
Light, Larry, 80–1
Lim, Victoria, 97–8
Limbaugh, Rush, 115
Liodice, Bob, 67
Local content and convergence, 73, 111–3, 121–2
LoCicero, Geoff, 94–5, 141, 143
Locin, Mitch, 41, 42, 45–6, 50, 117–8
Lukasiewicz, Mark, 140

M

Madigan, John, 118
Maeil Business Newspaper, South Korea, 128, 172, 215
Magnet locations, 215
Manchester Evening News, The, 32, 129–30, 145, 217–8
Marca, 6
Maxwell, Scott, 48
Mayne, Peter, 138
McAdams, Mindy, 52
McCabe, Joseph Jr., 35
McCartney, Robert, 120–1
McCombs, Regina, 152
McFarlin, Diane, 64–5
Media Center at API, 19
Media Corporation of Singapore, 38
Media General, 51, 78, 116
Mediamorphosis, 76
Mendenhall, Preston, 98–9, 144–5
Mentoring, 171
Miller, Stephen, 162
Mindset and convergence, 6, 15, 22, 63–6, 86–7, 131–2, 154, 157, 158–9, 162, 193, 205, 206
Ming Pao, 22–3, 46, 128–9
Mobile phones, 90
Moblogging, 92–3, 140, 141–4, 148, 197
Moor, Anthony, 53, 56, 58, 214
MORI, 82
Morningstar, 73
Moroney, Jim, 46
Morse code, 179
Moving the furniture, 216–7
MSNBC.com, 79, 99, 140, 144
MTV3 TV, Finland, 123
MUDIA consortium, 43, 219
Multi-media advertising and convergence, 70, 77–83
Multi-media as future, 205
Multi-media desk (aka "bridge"), 108, 116, 216
Multi-media reporting process, 147–8
Multi-media reporting skills, summary, 102–3
Multi-skilled reporter, 96–7
Multi-skilled reporters, profiles of, 97–9

Index

Musser, Rick, 210–11
Myles, Paul, 37–8

N

Newswall, 146, 218
Nathaniel, Naka, 98
Navarra University, Pamplona, 74
New media and convergence, 76
New York Times, The, 35, 120
News Center, The (Lawrence, KS), 58, 161
News Center, The (Tampa, FL), 53, 54, 56, 64, 219
News cycles, 24-hour, 204
NewsGate (software), 53, 218
Newspaper Association of America, 32, 205
Newsplex Convergence Monitor, 132
Newsplex, 6, 129–30, 145–50
Newsroom, 24-hour, 31, 204
Nichols, Julie, 131, 141
Nokia, 125, 150, 190
NYTimes.com, 105, 120
NewsGear, 101, 150–1, 193
New prime time, 195–6
Nielsen/NetRatings, 32–3
Newspaper cultures, 71–4
New reporting roles and convergence, 91–7
News flow manager, 92–3
News resourcer, 94–5, 103, 172
Nachison, Andrew, 4, 19, 40, 63, 188, 203–4, 214,
Northrup, Kerry, 14, 20, 21, 59, 63–4, 75, 85, 90, 91–7, 101, 131, 142, 147–8, 158, 203, 204, 205, 207, 211–2, 214, 215–6, 218, 219

O

OhMyNews.com, 184–5
Online News Association, 29
Opinion leaders, 209
Organic convergence, 109
Organizational strategy and convergence, 44–45
Orlando Sentinel, 5, 10, 17–8, 22, 44–5, 99, 100, 116, 117, 206, 213
Ownership convergence, 9, 117

P

Peers, Martin, 28
Pew Charitable Trusts, 4, 31
Pew Internet Report, 183
Pfeiffer, Andreas, 18–9
Pirker, Horst, 76
Pitts, Gordon, 7, 117
Platypus journalism, 10–1
Powell, Michael, 34, 39–40
Presenting skills, 102–3
Priorities and convergence, 202
Procter & Gamble, 81
Promotion and convergence, 65
Proximity and convergence, 57, 159
Pryor, Larry, 6, 7, 8, 15–6
Pulitzer Prizes, 119

Q

Quality content, need for, 16, 66, 156, 194, 196, 221
Quicklink, 36, 138

R

Ramirez, Pedro, 189
Readership Institute, 71
Reddick, Randy, 181
RedEye, 30
Redfern, Ron, 51, 53
Reporters and copyright payment, 122–3
Reporting and convergence, 86
Rheinische Post, 66
Rodriguez, Jose Luis, 55
Role of managers and convergence, 24–5
Rosenblum, Michael, 36–7
Russin, Joseph, 58, 120
Rutherford, James, 20–1

S

Saffo, Paul, 25, 195
Salaverria, Ramon, 74
Sambrook, Richard, 199
San Francisco Chronicle, 17
Sarasota Herald-Tribune, The, 64
Schmidt, Mogens, 163
Self-selected news, 196–7
Serious Magic, (software company), 131
Shared values and convergence, 52–3, 206
Shenk, David, 29, 155
Simons Jr., Dolph, 44, 114
Singapore Press Holdings, 38, 187
Singer, Jane, 48, 209–10
SMS as reporting tool, 99, 115, 161, 189
Sony PD150 and 170 cameras, 37
South Korea and technology, 184
Speed and ethics, 192
Sport and convergence, 112–3
St Petersburg Times, The, 66
Stencel, Mark, 69, 121
Stewart, Tom, 164
Stone, Martha, 2, 6, 42–3, 77, 78, 79, 81, 85, 110, 121, 144, 145, 209
Story builder, 93–4
Storytelling convergence, 12, 88–90
Strengths and weaknesses of media, 88–90
Structural convergence, 11
Sturm, John, 32
Sulzberger Jr., Arthur O., 1–2, 45, 183
Sun, The, 189
Sunday Missourian, The, 30
Superstation WGN (Chicago), 119, 211, 214
Sveiby, Karl-Erik, 156
Sydney Morning Herald, The, 53, 67
Symbols of convergence, 216

T

Tactical convergence, 9–10, 117
Talk-backs, 102
Tampa Tribune, 5, 23, 116
TBO.com, 119
Technology and convergence, 32–5, 144–5, 162–3
Technology and reporting milestones, 35, 180

Index

Telefax, 123–4
Telegraph boosts newspapers, 180–1
Telephone's impact on reporting, 181
Television via cell phone, 190
Thelen, Gil, 5, 20, 31, 40, 46, 48, 49, 53, 56, 62, 63, 65, 134, 160, 186, 202, 203, 215, 220
Thumb generation, 189
Tidblad, Bella, 55
Tillinghast, Charlie, 79
Times, The, 70
TiVO, 28, 32
Tomorrow's News video, 90–1, 93
Tompkins, Al, 151–2
Training and convergence, 108, 163, 198–9
Training, increased need for, 198–9, 212–3, 220
Treffi, 125
Tribune Company, The, 118–9, 215
Trust and convergence, 167–8
Turku Television, 124
Turun Sanomat Group, 25, 70, 121–6
TV journalists train print people, 121, 128, 213
Tyner, Howard, 20, 185–6

U

Uncertain regulatory frameworks, 187

V

Valjakka, Ari, 3, 25, 70, 72–3, 74, 122–3, 124–5, 134–5
Values and convergence, 202, 211
Values, communication of, 202, 206
Van Klaveren, Adrian, 36
Verklin, David, 80
Veronis Suhler Stevenson, 21
Veseling, Brian, 19–20, 159
Video journalism, 36, 38, 130–1, 140
Visicom, 124
Vision Korea, 128
Visual Communicator (software), 131, 148, 150, 151
Vorarlberger Online, 143–4

W

Wall Street Journal, The, 27, 30, 194
Washington Post, The, 120
Washington Post.com, 121
Water cooler conversations, 212
Weaver, Janet, 215
WFLA-TV, 23, 79
Wheeler, Keith, 10, 99–100, 117, 135, 211
Wilkinson, Earl, 2
Wireless Election Connection, 92–3, 141–4
Wireless fidelity (wi-fi), 139
WKMG Channel 6, 10
Woie, Oyvind, 99
Wolpe, Bruce, 68, 69

World Company, Kansas, 15, 44,
    51, 106–16, 195, 206, 216
Wright, Dean, 79
Wurtzel, Alan, 28

X

XML (eXtensible Markup
    Language), 11, 175, 218

Y

Youth appeal of blogs, 115
Youth newspapers, 30

Z

Zeglis, John, 35

www.ingramcontent.com/pod-product-compliance
Lightning Source LLC
Chambersburg PA
CBHW052015290426
44112CB00014B/2254